T0321924

A PRACTITIONER'S GUIDE TO

Effective Maritime and Port Security

A PRACTITIONER'S GUIDE TO
Effective Maritime and Port Security

Michael Edgerton

WILEY

A JOHN WILEY & SONS, INC., PUBLICATION

Library of Congress Cataloging-in-Publication Data:

Edgerton, Michael, 1961–
    A practitioner's guide to effective maritime and port security / Michael Edgerton.
        pages cm
    Includes index.
ISBN: 978-1-118-09991-9 (cloth)
    1. Harbors—Security measures. 2. Shipping—Security measures. 3. Maritime terrorism—
Prevention. 4. Harbors—United States—Security measures. 5. Shipping—United States—
Security measures. 6. Maritime terrorism—United States—Prevention. I. Title.
    VK203.E34 2013
    363.28'9—dc 23   2012048236

Printed in the United States of America.

10 9 8 7 6 5 4 3 2 1

To my family with love

Contents

PART ONE

The International Maritime Operating Environment

PART TWO

Threats to Ports and the Maritime Domain

PART FOUR

Principles for Effective Maritime and Port Security

APPENDICES

Introduction

This book is designed for practitioners as well as students of maritime security, maritime transportation, and international business. The book provides an overview and analysis of the current factors that affect the security of the maritime operating domain, where shipping, international politics, economics, crime, and terrorism intersect in ways that have far-reaching global impacts. The book also provides a critique of the current approaches to maritime and port security that have been implemented by nations, the private sector, and the international community. The critique is an overview of the current state of maritime and port security as well as an assessment of potential challenges or weaknesses. This book provides actionable recommendations to enhance security while facilitating trade and improving the resilience of companies and governments.

In policy discussions, most of the current critiques of homeland-security measures and programs focus on national-level policy and do not differentiate between modes of transportation or infrastructure domains. This book is specifically focused on the maritime-security environment and proposes measures and approaches that will make international trade and maritime transportation more secure while enhancing efficiencies that can reduce the cost of security for ship operators, customers, and governments.

As part of this approach, the book assesses the business case for security and provides recommendations that are consistent with a focus on efficient, yet effective, security measures. The book is written for the private sector or government practitioner, as well as academic and government policy makers. Because of the inherently international nature of maritime commerce, this book takes an international approach to possible maritime and port security solutions without bias toward or against any approach advocated by a particular nation or organization.

The book is divided into four major sections. The first is the context, which provides an overview of the basic components of the international maritime operating environment. These components include the multinational nature of shipping and multinational drivers in port operations, the criticality of ports from several perspectives, including geopolitics, ports as potential targets, and ports as potential conduits for illicit activity. The contextual section will also cover the issue of port connectivity to other modes of transportation and the intermodal nature and linkages of port and maritime security.

The next section assesses threats to ports in the maritime environment. These threats can take several forms, including state-sponsored threats, conventional state military threats, terrorism, criminal actions, embargo violations, and corruption and lack of transparency. Understanding threats is essential to being able to implement effective maritime security measures.

The third section of the book provides a critique of the current approaches to maritime port security. This critique will look at the United States, the European Union, and other international measures and approaches. Specifically, this section will identify the prism through which authorities make port security policy and some of the factors that affect policy-making and the establishment of current programs and their strengths and weaknesses.

The fourth and final section of the book lays out suggested principles for effective, truly risk-based maritime and port security. These principles focus on the need for security to be more than an objective but in addition a key enabler for legitimate maritime trade. The book reframes port and maritime security as a key component of a multidisciplinary system in which secure, resilient, and efficient trade is the objective.

ACKNOWLEDGMENTS

This book was made possible only through the excellent editorial guidance and assistance of Virginia Howe and the late Dodge Woodson as well as the outstanding graphics support of Ben Spear. I am indebted to my employer, Good Harbour International, for supporting my efforts. I also need to acknowledge those with whom I served in September 2011, especially Joe Coccia, Mike Ferullo, and Paul Kohl. While they may not agree with everything in this book, we struggled to implement security measures on the fly with little or no guidance. As a result, they taught me some of the key lessons found in the book. This book wouldn't be possible without the support of my colleagues in government and the private sector with whom I've been privileged to work, all of whom have in some way, contributed.

MICHAEL EDGERTON

Foreword

A century ago when America thought about its security, much of its attention focused on its harbors. A large component of the US Army was then a branch called Coastal Artillery. That organization operated an extensive network of large fortresses that guarded America's harbors. Today those forts are historical relics turned into parks. Yet America today is far more dependent upon what comes in through its ports than it was a century ago. Then America was largely self sufficient; today it relies upon "just in time delivery" of millions of containers. Freight ships and tankers are our lifeline to the world, carrying vastly more than the small fraction of trade that moves by air cargo. Despite that shift in the importance of what comes through our harbors and what moves on ships, until relatively recently maritime security did not figure importantly in America's national security agenda. Not any more.

Today America's security and that of many other nations is intimately entwined with maritime security and governments now are recognizing that. The priority given maritime security has increased significantly in the last decade. Maritime security in the twenty-first century, however, is not a matter of large stone forts on harbor islands. What does comprise modern maritime security is the subject of this timely and comprehensive volume by my friend and colleague Michael Edgerton.

Mike has personal experience with maritime security in the U.S. military and in Department of Homeland Security. Now as a private sector consultant, Mike has examined various approaches to security risk. He has been able to identify leading best practices that could be adopted for worldwide implementation. His methodology reflects globally accepted approaches to risk management and the need for security to contribute to the broader resilience of the maritime transportation system.

In his work on maritime security issues, Mike has developed a sophisticated approach to maritime security that recognizes the role of both government and the private sector. Much of his work has focused on the critical importance of implementing security measures that are complementary to the broader requirement to drive the continued operation and resilience of the maritime domain under all but the most severe threat conditions. This book analyzes the components of the maritime transportation-logistics system, reviews the security measures which have been put in place, and offers a fresh, comprehensive, yet flexible approach to managing maritime security risks without unduly restricting the need for rapid and efficient transportation of goods and people. It is a valuable contribution that governments, corporations, and maritime operators can all benefit by taking to heart its insights and recommendations.

RICHARD CLARKE

Richard A. Clarke is an internationally recognized expert on security, including homeland security, national security, cyber security, and counterterrorism.

Clarke served the last three Presidents as a senior White House Advisor. Over the course of an unprecedented 11 consecutive years of White House service, he held the titles of:

- Special Assistant to the President for Global Affairs
- National Coordinator for Security and Counterterrorism
- Special Advisor to the President for Cyber Security

Prior to his White House years, Clarke served for 19 years in the Pentagon, the Intelligence Community, and State Department. During the Reagan Administration, he was Deputy Assistant Secretary of State for Intelligence. During the Bush (41) Administration, he was Assistant Secretary of State for Political-Military Affairs and coordinated diplomatic efforts to support the 1990–1991 Gulf War and the subsequent security arrangements. In a Special Report by Foreign Policy Magazine, Clarke was chosen as one of The Top 100 Global Thinkers of 2010.

PART ONE

The International Maritime Operating Environment

CHAPTER 1

Unique Characteristics of Ports and International Shipping

INTRODUCTION

International shipping operates in one of the most lightly regulated domains in the world. Huge ships staffed by a minimal number of crewmembers transit the high seas, which are outside of the jurisdiction of any country, and often fly the flags of countries that are largely unable to exert jurisdiction or protect the vessels that are their legal territory. There are about 40,000 ships that engage in international trade. These ships provide over 90 percent of imports to North America and carry about 80 percent of trade worldwide. The total volume of global trade is expected to double over the next two decades, with a particular focus on containerized cargo.

Large-scale commercial maritime trade is characterized by its multinational nature. Over the last 25 to 30 years maritime shipping has become increasingly international in scope and composition. Today, it is not unusual to find a ship that is registered in a country that it has never visited (or cannot visit, as it may not have a coast, such as Luxembourg or Mongolia), owned by a company whose representatives have never set foot on the ship, and operated by an additional company located in a country unrelated to the country of registration or country of ownership. It would also not be unusual for that same ship to have crewmembers with potentially questionable or unverifiable professional qualifications from several different countries, none of which are from the country of registry. These crewmembers would most likely have been hired through potentially unregulated and corrupt personnel agencies in their home countries. Further, because of the multinational nature of shipping, the maritime security environment throughout the world is inherently interlinked in ways that would not be apparent to those not familiar with shipping or maritime trade. This global interdependence makes security measures in countries that have direct or even indirect trade links with other countries important to security and border agencies, since they reflect the integrity of security on a vessel coming to the nation's port. An example of this is trade between Indonesia and the United States. There are vessels that trade in coal that transit directly

FIGURE 1.1

Foreign ships create temporary borders between countries that normally don't share a border.

between Indonesian courts and ports on the US East Coast. Therefore, the security on the ship is only as good as the security in the Indonesian ports at which it had previously visited. When that ship docks at a US port, there is essentially a temporary land border between Indonesia and the United States, and the effectiveness of Indonesian border and port security is directly and critically important to the United States.

The inherent interdependencies between national and international security regimes will be exacerbated as global maritime trade continues to grow. Despite the recession of 2008 and 2009, analysts expect a rapid increase in maritime trade to continue, perhaps doubling within the next 20 years. This increase, coupled with a philosophy of "just in time delivery," will place additional stressors on ports involved in maritime trade. The stressors will be a result of a need for delivery to meet timely, accurate schedules as well as aging intermodal infrastructure in which ports are located in congested urban areas where there is limited room for expansion and already crowded roadways and rail tracks. The "just in time delivery" approach is a relatively recent phenomenon that was a result of the need to reduce the costs of storage and warehousing. The result, however, is that transportation delays, regardless of the cause, can be catastrophic.

When studying approaches to maritime and port security, an additional challenge lies in trying to determine what constitutes a port. Unlike airports, which developed more recently and are more likely to have clearly defined borders and less likely to be found in the older, more crowded parts of cities, ports are often hundreds of years old and grew up within cities—or in many cases cities grew up around ports. As a result, port functions often occur outside the formal defined perimeter of a port. These functions include warehousing, the operation of free zones, transshipment points, providers of maintenance and supply services, shipping agents, and shipping-company headquarters. Further, intermodal connections may not be within the formal perimeter ports. Rail hubs, container yards, and trucking centers may also be remote from the port. These issues complicate attempts to understand what constitutes a port as well as jurisdictional boundaries for ports and the maritime domain.

THE MULTINATIONAL NATURE OF SHIPPING AND BUSINESS DRIVERS IN PORT OPERATIONS

Ship and port operations are complex and involve numerous industry and government organizations and entities. These include:

- Vessel registries
- Ship owners
- Ship operators
- Classification societies
- Personnel agencies
- Labor unions
- Shipowner/operator organizations
- Shipping agents
- Cargo brokers
- Third-party logistics carriers
- Port operators
- Terminals
- Port authorities
- Harbormasters
- Customs
- Immigration agencies
- Coast guards
- Police agencies
- Intelligence agencies
- Navies
- Trade associations
- Safety agencies
- Environmental agencies
- Charterers
- Insurers
- Ship chandlers
- Technical repair and maintenance companies

International shipping has three larger components into which all of the above entities fit. These components are:

- Flag states—those countries in which ships engaged in international trade are registered
- Port states—those countries in which the ports that ships engaged in international trade visit are located
- Supply chain—the system in which goods and materials are transported form point of origin to point of use, often involving more than one mode of transportation

FLAG STATES

Flag states have unique responsibilities because vessels registered under the flag of a country form, for legal purposes, a part of that country's territory that moves around the world. Therefore, the flag state has responsibilities for the safe and secure conduct of ships registered under its flag, regardless of where the ship is. This is complicated by the fact that vessels on international voyages also enter the jurisdiction of port states, thereby making them subject to some port-state rules as well.

Vessel Registries

The multinational nature of shipping is most apparent in how ships are owned, staffed, and operated. At the forefront of this are vessel registries. A vessel registry provides the mechanism for flag states to manage their commercial maritime fleet. Through a registry, a nation can create and enforce rules and requirements for vessels that fly its flag. Further, flag states have the primary responsibility for ensuring that vessels within their registry meet national and international requirements to which the flag state is signatory. Some of the functions of a flag state include:

- Conducting safety inspections
- Monitoring vessel compliance with all international and national standards, including security
- Investigating accidents and misconduct
- Issuing certificates of registry, staffing documents, and other required documentation
- Overseeing seafarers' licensing and documentation

Types of Vessel Registries

There are three basic types of vessel registries in international shipping:

- Traditional national registries
- Second registries
- Open registries

Traditional National Registries

The traditional national vessel registry involves owners, operators, and crews from the same country in which the ship is registered. Until World War II, this approach to flag registries was used almost exclusively. The system guaranteed employment for large numbers of citizens and ensured that vessels registered in flag states would be fully regulated under that state's laws. Ships registered under conventional flag-state registry were able to request naval protection from that flag state anywhere in the world where that nation's naval units operated. Examples of conventional or traditional registries that still exist include the United States, Canada, and, to a lesser extent, some European Union member states.

Second Registries

With the rise of open registries (flags of convenience), conventional and traditional flag states have created additional registries, which they term "second registries." These registries have some of the characteristics of the conventional flag-state registry but also include some of the features of an open registry. For example, second registries will often adhere to the regulatory requirements of a conventional flag state but will allow crewmembers who are not nationals of the flag state to serve on board, including in command positions. This reduces labor costs, which are traditionally one of the higher costs involved in ship operations.

Open Registries

Since World War II, crews from developed countries have become increasingly expensive, and shipowners have perceived safety and pollution-prevention regulations to become more strictly enforced in the developed world. Therefore, in order to maximize profits, shipowners often pursue vessel registrations that are cheaper and have less regulatory requirements than the vessel registries of more

FIGURE 1.2

Mongolian flagged ship.

established nations. These registries are known as open registries or "flags of convenience." The first open registry was created in Panama in the late 1920s. Open registries became more widespread after the Second World War, especially after the creation of the Liberian open registry in the late 1940s.

Open registries are usually found in smaller, less developed countries and are used as a way to raise revenue while providing ship operators and owners with an incentive to utilize their services by keeping regulation light and costs down. Today, the largest open registries are Panama, Liberia, and the Marshall Islands. In these and other cases, open registries are often administered outside the formal political or governmental administration of the flag state. For example, the Marshall Islands flag Registry is managed in Reston, Virginia, by a private company, International Registries, Incorporated, on behalf of the Marshall Islands government.[1] The following is a non-exhaustive list of countries that operate open registries and have a limited international presence and capability to enforce laws and regulations outside their territory:

- Vanuatu
- Malta

- Tuvalu
- North Korea
- Belize
- Cambodia

Further, some landlocked countries such as Mongolia operate open vessel registries even though it is clearly impossible for them to provide naval support. Their efforts to oversee the enforcement of regulations on ships flying their flag is also limited by the requirement to outsource their maritime oversight, since they have no pool of maritime professionals to engage.

The advantages of open registries include: the ability to hire mariners from countries in which labor costs are lower, lower fees and tax burdens, the ability to reduce the transparency of ownership, and less stringent regulatory requirements. In many open registries it is possible for the identity of shipowners to be protected.

The UK's Committee of Inquiry into Shipping prepared a report in 1970, which provides a set of characteristics typically found in open registries,[2] including:

- Registration by foreign citizens is permitted (and in fact, encouraged).
- Minimal to nonexistent taxes
- The registry country is small, and tonnage charges may produce substantial effects on national income
- Ship operators can hire third-country labor; the flag state has no requirement for its nationals to be hired, even in command positions
- The flag state in incapable and unwilling to impose domestic or international regulations

All of these characteristics of open registries are attractive to shipowners and operators, who see light regulation and taxation as vital to the reduction of operating costs and the maximization of profit.

Implications for Security

The prevalence of open registries in the last few decades has resulted in less government oversight into shipping both operationally and administratively. As a result, the tendency towards open registries has increased the multinational nature of shipping, arguably decreased the amount of government oversight of shipping activities, and decreased the transparency of vessel ownership and operators, as open registries have less stringent regulatory and financial reporting requirements than conventional flag-state registries. In open registries, shipowners often create shell companies that disguise the actual shipowner or shipowners. Concealed ownership is advantageous to entities that are trying to minimize tax burdens, avoid taxes, or are involved in illicit or potentially illicit activities.

Third Country Owners

Vessel owners do not necessarily have any political, social, or cultural links to the flag states that operate open registries. A key attraction for owners to register their vessels under flags of convenience is the ability to make ownership hard to trace. This opaqueness is under pressure as a result of post-9/11 interest in maritime security and the potential threat of shipping as a terrorist conveyance. However, the intricacies of ship financing and the combination of international interests and light governance of shipping have led to the retention of an opaque ownership structure among open registries.

The Organisation for Economic Cooperation and Development's (OECD) Directorate for Science, Technology, and Industry prepared a study in 2003 entitled "Ownership and Control of Ships," which identified several key findings regarding ship ownership.

The report essentially finds that it is easy and inexpensive to establish a web of corporate entities and instruments, such as limited-liability companies, nominee shareholders and directors, and bearer shares to provide very effective cover to the identities of owners who do not want to be known.[4]

Additionally, open registries, which have very limited or no nationality requirements, are easy jurisdictions in which to register vessels that are covered by complex legal and corporate arrangements. The multinational nature of shipping means that the corporate arrangements will involve a number of nations, thus complicating any attempts to determine beneficial ownership.

Implications for Security

The lack of transparency of ownership found in open registries (although not limited to open registries) and lack of interest on the

CASE STUDY

In late 2002, the *So San* a North Korean ship sailing under Cambodian registry, was discovered with 15 Scud missiles and 15 conventional warheads, 23 tanks of nitric-acid rocket propellant, and 85 drums of unidentified chemicals hidden beneath a cargo of cement bags. The weaponry had documentation indicating it was headed to Yemen but ended up being sent to Libya. The Cambodian government has taken a very "hands-off" approach on its responsibilities as a flag state, as indicated by Ahamd Yahya of the Cambodian Ministry of Public Works and Transport, who reportedly said, "We don't know or care who owns the ships or whether they're doing 'white' or 'black' business … it is not our concern."[3]

FIGURE 1.3

M/V *So San* being boarded by Spanish special forces. *Photo credit: US Navy.*

part of some registries to encourage or require transparency, coupled with the lack of strong international regulation and disclosure requirements for vessel ownership, present several security challenges, including the ability to profit from legal and illegal trade anonymously as well as the ability to minimize governmental involvement in illegal activity.

Vessels owned by entities or individuals that have no direct link to the operator or largely unregulated flag registry may more easily be used for illicit activity, as there is no strong governmental control over the vessel and the true operational control of the vessel may be hidden from port-state authorities who are responsible for assessing potential high-threat vessels that may be arriving in their ports. Under the current system, dishonest owners are likely to have a high degree of confidence that the risks they incur for smuggling, operating unsafe vessels, or other illegal activity are acceptable and that their involvement will not be discovered, even if the ship is seized or their activity is disrupted. Further, the lack of transparent ownership information complicates the work of government investigators, who may be pursuing criminal cases or developing intelligence leads. The current system, which discourages the transparent and open identification of beneficial owners of vessels, is likely to increase the security risks throughout the global shipping industry as well as in ports of call.

Multinational Crews

Crews from several nations serving on the same vessel bring additional potential security challenges. These challenges include:

- Disparate levels of training, qualification, and overall professionalism depending on crew nationality
- Possible intracrew hostilities or tensions based on nationality
- Potential lack of crew cohesiveness
- Crew lack of familiarity with each other over long periods of time

In recent decades, the composition of merchant crews has changed significantly, with a greater percentage of crews coming from developing countries, some of which have less technologically advanced methods of documenting or validating crewmember identities and qualifications.

Crew costs are one of the highest expenses that ship operators incur, comprising up to 50 percent of vessel operating costs. In the late 1990s an American ship's master could expect to earn about $11,000 per month, while a Filipino ship captain would earn about $2,000. Therefore, considerable savings can be realized by hiring merchant mariners from developing countries at salaries that are a fraction of the cost of hiring crew from North America or Western Europe. While mariners from developing nations may be competent, the risk of using unqualified personnel increases due to the potential for crewmembers to have received poor training in "diploma mills" or potentially paying off corrupt officials or staffing agencies for work. These issues have been documented by international shipping organizations in the past, although recent pressure has reduced some of the undesirable activity.

Implications for Security

Most mariners who are not ship's officers (deck officers or engineering officers) are not permanent employees of the ship operator on whose ship they are working. They are typically employed for specific trips or periods of time through staffing agencies that procure labor for ship operators. In many cases, there is evidence of crewmembers paying up to one month's salary to the agency for the opportunity to work. Officially, the fees for the agency are supposed to be paid by the ship operators. The illegal requirement for a mariner to pay part of his salary contributes to a culture in which mariners' documents and qualifications are less intensely scrutinized.

PORT STATES

The term, "port state" is used to identify those nations that have commercial maritime ports that are subject to regulatory oversight by the national government. The term is used primarily for those activities that govern the oversight of foreign ships in a nation's territorial waters. The term is also specifically relevant, as it specifies the duties of nations signatory to international codes, treaties, and other agreements. There is no correlation between the importance of seaports

CASE STUDY

In the late 1990s, port-state regulators in North America and Europe became increasingly concerned by the prevalence of qualification certificates being issued to Filipino mariners by training institutions of questionable credibility. These training institutions were labeled as "diploma mills," where mariners would get rudimentary training (or, in some cases, no training) in their specialty but would receive a diploma that indicated they were fully qualified or licensed. While the Philippines was not the only country with limited regulation of maritime training, Filipino mariners comprised such a large percentage of merchant mariners throughout the world that the problem received considerable attention at the International Maritime Organization. The resulting international pressure led the Philippine government to close down the diploma mills and provide a government-sponsored web portal to verify mariner documents issued by the Philippines.[5]

and the level of robustness of the country's fleet. Many of the countries where the world's largest seaports are located have small fleets, and many of the largest fleets in the world are registered in countries with small or no seaports. For example, the five largest flag registries and the five largest maritime trading countries are largely mutually exclusive, as noted below.

The five largest flag registries are[6]:

- Panama
- Liberia
- Marshall Islands
- Hong Kong[7]
- Greece

In 2008, the five largest cargo ports, by cargo volume, were[8]:

- Singapore
- Shanghai, China

LARGEST FLAG REGISTRIES	LARGEST CARGO PORTS
Panama	Singapore
Liberia	Shanghai, China
Marshall Islands	Rotterdam, Netherlands
Hong Kong	Tianjin, China
Greece	Nongbo, China

FIGURE 1.4

There is no significant correlation between the largest
fleets of ships and the largest ports.

- Rotterdam, Netherlands
- Tianjin, China
- Ningbo, China

Regulatory Requirements

Even though ships registered in different countries have legal status
as territory of the flag state, the fact that they move around the world
and enter the territorial seas of port states means that port states are
able to impose regulatory requirements on those ships and enforce
international requirements to which the flag state is signatory. Port
states vary in the level of regulatory enforcement they exercise over
foreign vessels.

Countries in the developed world tend to have well-developed
port-state regulatory frameworks that allow for the enforcement of
international treaties and codes and national regulations pertinent
to security.

International Treaties and Codes

Port states have a recognized set of rights and responsibilities in exer-
cising limited safety and security oversight of foreign vessels transiting a state's territorial waters or visiting its ports. Historically, port-
state responsibilities were focused on ensuring that vessels adhered to

international pollution-prevention regulations or vessel safety standards. After the attacks of September 11, however, port-state oversight included new, security-related treaties, codes, and conventions. Some of the key international treaties, codes, or conventions that influence or inform port-state control efforts include:

- 1982 United Nations Convention on the Law of the Sea (LOSC): LOSC addresses a full spectrum of legal issues pertaining to the seas and is applicable to both port and flag states. It is generally viewed as the most universal fundamental statement of international law in virtually all matters pertaining to the seas, even for states that are not party to it. LOSC addresses maritime issues by category, such as the territorial sea, straits, the continental shelf, the high seas, environmental protection, deep seabed mining, regional cooperation, dispute resolution, and more. Further, certain sections set out the rights and obligations of port states, balanced by the right of shipping to engage in "innocent passage." Of note, the United States, while signatory, has not ratified the Convention but adheres to its requirements.

- 1948 International Maritime Organization Convention (IMOC): The IMOC established the International Maritime Organization (IMO) as a specialized agency of the United Nations, specifically focused on maritime issues. The IMO's responsibilities include marine safety, marine environmental protection, maritime security, and maritime legal regimes. The IMO is accepted as "the competent international organization" referred to in the LOSC.

- 1974 International Convention for the Safety of Life at Sea and its Protocol of 1978 (SOLAS 74/78): SOLAS is the primary convention that governs maritime safety and security and forms the basis for many port-state regulations. It includes international standards in areas such as lifesaving requirements, navigational safety, crew licensing and competence, and vessel management. The 2002 International Ship and Port Security (ISPS) Code is included in the convention and constitutes the primary standard for maritime security for both ships and ports. It includes requirements for vessel and port facility security plans, designated security officers, exercises and training, advance arrival notice, and more. It also requires those plans to be approved by a national or other authority, acting on a country's behalf; and to implement the measures in those plans. Because it is part of SOLAS, the ISPS Code is widely observed. The convention also provides the legal authority for the port state to inspect, investigate, and possibly detain foreign ships that are not compliant with the required security standards.

- 1988 Convention for the Suppression of Unlawful Acts Against the Safety of Maritime Navigation (SUA): SUA 1988 has two major elements. First, it extends the jurisdiction of a country over violent acts committed on vessels on the basis of vessel flag, location, or nationality of either offender or victim. SUA 1988 applies to any international voyage. Additionally, it requires countries with established jurisdiction to either prosecute the suspected offenders or extradite them to another location for prosecution. As a result of the September 11, 2001, terrorist attacks, additional protocols were added to include other potentially terrorist-related offenses such as using a ship as a weapon or the transportation of terrorists, weapons of mass destruction, or related cargo, including components or precursors. It also establishes procedures for port states or third countries on the high seas and boarding of vessels in the case of reasonable suspicion without the express consent of the flag state in certain situations. This convention also has applicability in counter-narcotics and counterpiracy operations.

Oversight Mechanisms

Oversight of the implementation and adherence to international codes and conventions is typically driven by the implementation of national legislation or regulations that codify the international conventions as national law. In some cases, the national laws and regulations expand the international legal frameworks for domestic application and create greater regulatory requirements than those in the international convention. International treaties and conventions have to be, by their nature, somewhat general, as they are a reflection of a collaborative process often involving dozens of countries.

The United Kingdom, Australia, European Union, and United States are among the states or entities that have developed sophisticated internal legislation to enhance the performance of port-state security responsibilities. Some of these national regulations and legislative instruments will be examined in greater detail in future sections of this book:

- European Community Regulation No 725/2005 (EC 725)

- European Union Directive 2005/65/ (EU 65).

- UK Statutory Instrument No. 1495: 2004: The Ship and Port Facility (Security) Regulations 2004

- 2002 US Marine Transportation Security Act (US MTSA) and corresponding Title 33, US Code of Federal Regulations, Part 101 (33 CFR 101)

- Australian Maritime Transport and Offshore Facilities Security Act 2003 (Australian MarSec Act)

Ship-Port Relationships

A microcosm of the relationship between a flag state and port state is most clearly highlighted in the relationship between a foreign ship that enters a port to load or unload cargo. In an event repeated hundreds of times a day, a vessel flying the flag of a nation enters the territory of another nation to deliver or load cargo or passengers. The vessel is the legal territory of a country that likely does not share a land border with the country it is visiting. However, for the period of the ship's visit, the flag state and port state share what is essentially a mobile, temporary border, with many of the same challenges, including potential threats to the ship and the port as well as the need to cooperate to ensure the smooth flow of commerce and trade. Therefore, although both foreign ships and port states retain their respective rights and responsibilities, the inherent requirement to cooperate in ensuring both ship and port safety and security is reflected in the ISPS Code by the requirement that in any non-routine situation, the ship and cognizant port authority execute a Declaration of Security, which outlines the safety and security measures to be taken by each party and the appropriate emergency contact information.

The Supply Chain

Ports and ships ultimately form a component of a broader supply chain, which includes the movement of goods, people, and information between points of origin and their final destinations. Supply chains can be very complex, with many different components, or fairly simple. Regardless of the complexity, global industries and national economies have become increasingly dependent on supply-chain models that are responsive to rapidly changing requirements. This is largely due to a decrease in the ability or desire for companies to retain material or products in large warehouses for extended periods of time and an increasing reliance on "just in time delivery."

Just-in-Time Delivery

Just-in-time delivery has become more important in recent years as companies, many of which rely on international suppliers for parts

----

FIGURE 1.5

Trucks waiting to cross the US–Canadian border in September 2001. *Material reprinted with the express permission of Windsor Star, a division of Postmedia Network Inc.*

CASE STUDY

In the hours after the attacks of September 11, 2001, the border between the United States and Canada was shut down. Among other impacts, trucks carrying auto parts manufactured in Ontario were unable to make their deliveries to auto-assembly plants in neighboring Michigan. This resulted in Ford closing five plants in Michigan for several days and sustaining a 13 percent decrease in vehicle production for the quarter.[9]

or products, try to reduce costs by minimizing storage requirements. The result is an increased need for reliable delivery of their supplies with minimal disruption. Any disruptions could prove to be disastrous to a company, region, or even a country.

The Components of a Maritime Supply Chain

Maritime supply chains can be extremely complex or relatively simple. In all cases, there are basic components to international supply chains:

- The producers of goods or material—this may be a factory, farm, or other point of origin
- Freight forwarders or consolidators—specialists in preparing cargo for transportation and insertion into the supply chain
- Surface-transport operators—truck or train operators
- Customs—government officials charged with collecting duties on goods and enforcing laws on importation and exportation
- Port operators
- Ship
- Customs brokers—specialists in coordinating and facilitating the approval of goods to move through customs

- Buyers
- Point of delivery

In complex supply chains involving several countries, there may be multiple sets of these elements. An example of a supply chain in which the point of origin and the point of delivery are in different countries is presented in outline in Figure 1.6.

Regulatory Issues

Regulatory issues governing the supply chain are similar to port security regulations, as they rely on international cooperation, codes, and treaties to create a globally accepted system of security and oversight for the elements of the supply chain and goods moving through the supply chain. Due to the nature of supply-chain operations, regulatory oversight is complex, and there is a constant tension between the facilitation of legitimate trade and the enforcement of international and national laws and regulations. Some of the key regulations governing supply-chain security include a framework developed by the World Customs Organization as well as US and European Union supply-chain regulations:

- 2005 Framework of Standards to Secure and Facilitate Global Trade (SAFE Framework): 145 of the 171 WCO member countries have committed to implementing the World Customs Organization SAFE Framework. The SAFE Framework focuses on four core principles in an effort to ensure effective supply-chain security while facilitating trade: harmonizing electronic cargo information requirements, applying a consistent risk-management approach to security threats, using non-intrusive detection equipment to examine high-risk containers and cargo, and providing benefits to businesses who comply with the Framework's standards and best practices. Businesses include manufacturers, importers, exporters, terminal operators, and virtually every element in the supply chain. Those who show compliance will benefit by having their goods move more quickly and with fewer examinations.
- European Union Customs Security Programme (EU-CSP): Like the SAFE Framework, EU-CSP is a voluntary program

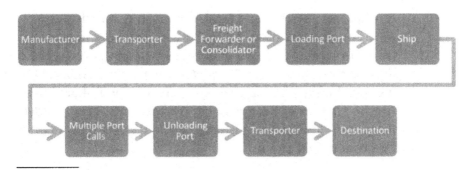

FIGURE 1.6

A supply-chain flow chart.

open to traders operating within the European Union. Member states will be entitled to grant expedited customs clearance to any business that meets established criteria relating to the operators' security control systems, financial solvency, and compliance record.

Within the United States, there are cargo and supply chain security programs across federal government agencies that are a result of legislation or other requirements. Most of these programs are geared towards the importation of goods, although there are programs that regulate the export of goods and services.

- Customs-Trade Partnership Against Terrorism (C-TPAT): C-TPAT is a voluntary, incentive-based program managed by Customs and Border Protection (CBP) in which participants agree to follow established program guidelines. Participants are expected to maintain a specified level of security by adhering to those guidelines and may be audited by CBP. As an incentive, participants receive priority processing at the border and training opportunities.

- Container Security Initiative (CSI): The CSI program, also managed by CBP, involves bilateral agreements between the US and foreign governments for over 50 international ports from which goods are shipped to the US. CSI allows for the assignment of CBP personnel at the foreign ports of loading. These CBP personnel identify high-risk containers bound for the US, which, if agreed to by host customs officials, are then screened (using x-ray and radiation-detection equipment) prior to loading. Once screened under the CSI program, the containers are typically not screened upon arrival in the US. The *quid pro quo* in the CSI arrangement is the expectation that CBP personnel will share intelligence with their host customs counterparts. CSI includes provisions for foreign customs officials to be stationed at US ports if they request.

There are several other international, foreign, and US programs designed to enhance the security of the global supply chain. Several of them will be examined in more detail in following sections of this book.

Intermodal Links

The global supply chain includes all forms of transportation and processes that can move cargo or people. The associated transportation conveyances can be placed in one of three categories; surface, aviation,

---

FIGURE 1.7

Intermodal links within a supply chain.

or maritime. Surface transportation consists of truck, rail, and mass transit transportation, while maritime transportation includes commercial shipping, passenger vessels, and ferries. Aviation includes air cargo and passenger transportation.

Any maritime transportation element within a supply chain almost always has an interface or connection with another mode of transportation. For example, goods produced in a factory may be transported by truck to a rail station, loaded on a freight train which transports the goods to a seaport, where it is loaded on a ship. When the ship docks, the goods will be unloaded and taken to their final destination by land transport. At each of these parts of the voyage, when the cargo is transferred from one mode of transportation to another, it passes through an intermodal link.

Key vulnerabilities in the supply chain are magnified when goods or people transit through intermodal links between modes of transportation, because government agencies with jurisdiction in one area may not have authority over other elements of the supply chain or infrastructure in another area. This may contribute to incomplete oversight of the potential security threats as they pass through the different modes of transportation on their way to their final destination. Because of the lack of comprehensive oversight of the supply chain, security regulatory measures often vary significantly between modes of transportation. While several of the previously identified codes and programs have measures that can reduce vulnerabilities in the supply chain, they are voluntary and therefore, inconsistently applied.

REFERENCES

1. "About IRI," International Registries, Inc., http://www.register-iri.com/index.cfm?action=about, accessed on 2 April, 2011.
2. Committee of Inquiry into Shipping (1970): *Report* (Rochdale Report). Cmd 4337, HM Stationary Office, London.
3. Ship Registers Feature: "Kings, Communists and Pushers: The Strange Ways of the Cambodian Register," *Fairplay International Shipping Weekly*, 12 October 2000.
4. "Beneficial owners" is a term that refers to the actual owner, whether it be an individual or corporation, that is the prime beneficiary of profits and benefits accrued from vessel ownership. This term is used to define the actual owners of vessels and other property whose ownership may be hidden by a series of companies and other legal instruments.
5. "Verification of Professional Licenses," Republic of Philippines Professional Regulation Commission, http://www.prc.gov.ph/services/default.aspx?id=16, accessed 10 July 2011.
6. United Nations Conference on Trade and Development, *Review of Maritime Transport: 2009*. New York: United Nations, 2009, p. 55.

7. Hong Kong's flag registry is operated independently of the People's Republic of China's registry due to its status as a semi-autonomous Special Administrative Region.
8. "World Port Rankings 2008, American Association of Port Authorities, http://aapa.files.cms-plus.com/Statistics/WORLD%20PORT%20RANKINGS%2020081.pdf, accessed 9 July 2011.
9. "Just-in-Time to Just-in-Case," http://www.gscreview.com/jan09_jtjust_n_case.php, accessed on 9 July 2011.

CHAPTER 2

The Criticality of Ports: Why and How They Matter

INTRODUCTION

To assess the criticality of ports to global maritime trade, national economies, and links to other modes of transportation, a full understanding of trade routes and chokepoints must be combined with a strong understanding of geopolitical considerations that can influence or drive changes in these trade routes. Further, ports need to be considered from two distinct perspectives when assessing criticality: ports as potential targets of illegal activity that may impact their ability to function as intended, and ports as conduits into and out of national borders and supply chains, which can be exploited in order to introduce or move illegal materials, persons, or activities. This understanding is necessary in order to develop a truly risk-based approach to managing the extraordinarily complex problem of maritime security, especially when coupled with limited resources to perform security missions and a need to facilitate lawful trade.

GEOPOLITICAL CONSIDERATIONS

Trade Routes

Unlike airports or rail terminals, ports are often hundreds or thousands of years old. In many cases, they are located in areas with natural harbors in cities that grew around them and are surrounded by very dense populations. The routes between ports form a global network of trade paths that are, for the most part, informally recognized routes that comprise the shortest distances between places that engage in trade with each other. These trade routes are not static and are subject to change as ports alter the commodities or products they trade in, security issues change along the routes, or the cost of shipping affects decisions regarding the most efficient routes. Further, trade routes in some cases do not follow patterns that are particularly intuitive and may be based on variables that are not immediately apparent. These variables and the economics of international trade can create efficient trade routes that do not conform to conventional understandings of geography.

For example, in recent years the Port of New York and New Jersey has seen an increase in shipments of containers from Asia and has therefore sometimes been labeled as a Pacific Rim port not because of conventional geography but because of the development of an expanded trade route. This is due to a decrease in labor costs that has made the shipment of containers by sea more cost-effective than shipping by road or rail from US West Coast ports.[1]

Trade Chokepoints

Throughout the world, there are areas, primarily between adjoining bodies of water that are relatively narrow, that are easy to monitor and, if necessary, blockade. In some cases, they are naturally occurring, and in others they are canals or channels created specifically to enhance the efficiency of trade by providing new and shorter trade routes. Because of their vulnerability to blockade, they are often at the center of geopolitical debates and negotiations and in some cases are subject to elaborate international treaties or agreements. The geopolitical security of chokepoints is an additional factor that can affect the stability of trade routes. Some of the world's most significant chokepoints include:

- Bab Al-Mandeb
- Malacca Straits
- Pillars of Hercules
- Bosphorus Straits

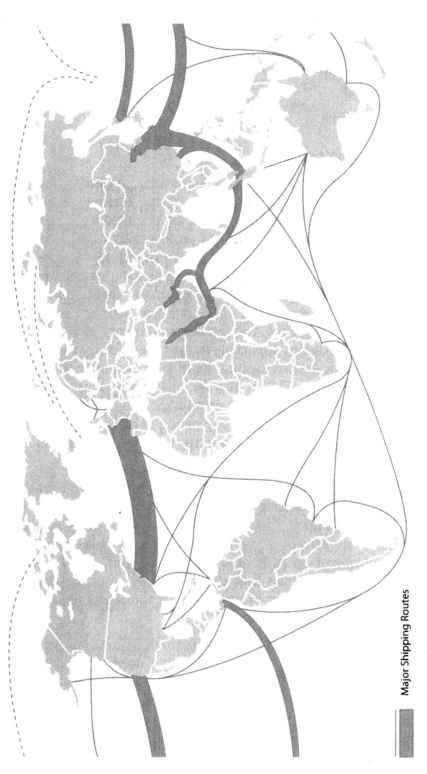

Major Shipping Routes

- - - - - **Projected Shipping Routes**

FIGURE 2.1

Global trade routes. *Map credit: UNEP/GRID-Arendal: http://www.grida.no/graphicslib/detail/the-boom-in-shipping-trade_1667.*

- Hormuz Straits
- Suez Canal
- Panama Canal

Sea Lines of Communication

Both maritime trade routes and the chokepoints through which some of the trade routes flow comprise what naval strategists term sea lines of communication (SLOCs). This term, while not invented by the US naval strategist Alfred Thayer Mahan, was promoted by him in his important work *The Influence of Sea Power Upon History* as a fundamental reason for the US Navy to expand into a global presence. He posited that ensuring the viability of SLOCs was in the vital national interest of the United States, as most of US trade was borne on ships in international trade. Therefore, the fundamental theory of SLOCs and their importance remains applicable to maritime security today, for many nations particularly regarding the threat posed by modern piracy and nonstate actors. SLOCs may be vitally important, even among nations in economic, political, or military competition, and can rarely be considered vital for only one nation or bloc of nations. An example of this complexity is found in the trade routes that pass through the Straits of Hormuz.

The Straits of Hormuz form a chokepoint through which over 30 percent of the world's seaborne oil is moved.[2] Key consumers of oil coming through the Straits include Japan, China, and the US. While China and the US are perceived as international rivals, their economies are interdependent on each other in many sectors, and they both need oil from the Arabian Gulf region, thereby ensuring that they both have a vested interest in ensuring that the Straits remain open for shipping.

In recent years, the Iranian government, from which China gets most of its oil in the Gulf, has threatened to shut down the Straits in the event of hostile action against it or its interests. Iran has also held military exercises that highlight its capability to close the Straits. However, as Iran has significant trade in oil as well as other products with China and other nations, any closure of the Straits could be self-damaging due to loss of trade income. Further, an Iranian closure of the straits would significantly threaten China's energy supply and

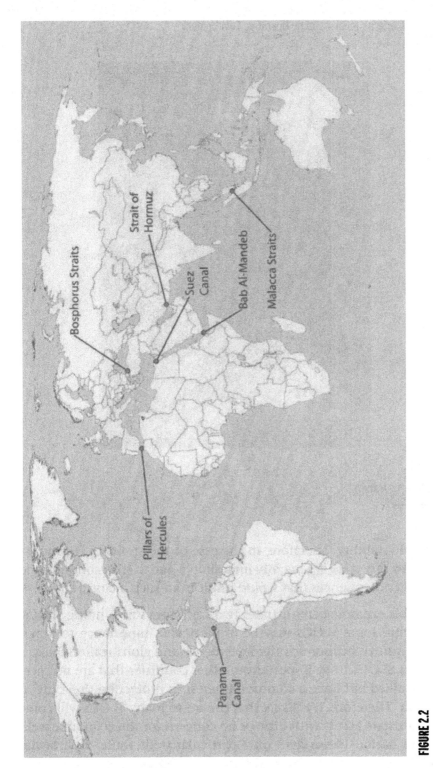

FIGURE 2.2

Maritime chokepoints.

FIGURE 2.3

Alfred Thayer Mahan.

economic stability. Therefore, in the case of the Straits of Hormuz, a closure by any state is generally undesirable due to the economic and political consequences that would result from such a closure.

This example demonstrates two key issues surrounding SLOCs. First, while some SLOCs may be of particular importance to a particular country, economic interdependence and globalization make it likely for SLOCs to be important to other countries that are not necessarily allied but have a common interest in protecting the route in question. Therefore, it is likely that threats to many SLOCs will come from nonstate actors with little or no concern for the economic well-being of nations dependent on a particular trade route, a particular set of ports, or a particular maritime chokepoint.

PORTS

When analyzing the criticality of seaports from a maritime security perspective, they should be considered as both potential targets for illicit activity or attack and as potential conduits for the illegal movement of cargo and people. This dual role increases the number and complexity of the vulnerabilities to which seaports are potentially subject. Further, the criticality of seaports is two-dimensional. Ports are critical due to their roles as the termini of the bulk of international

FIGURE 2.4

Straits of Hormuz.

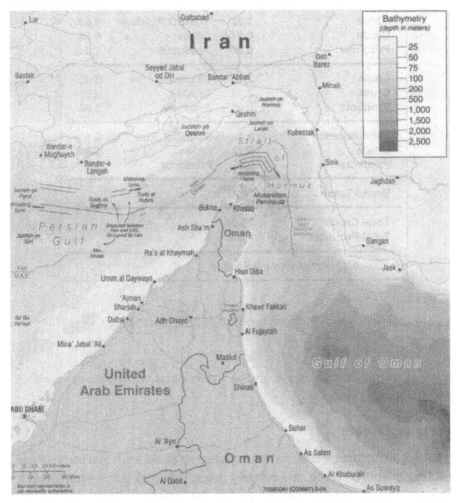

Destination of Exports Passing Through the Straits of Hormuz - January to October 2011 (million barrels per day)			
OECD Europe	0.83	Middle East	0.05
Crude	0.70	Crude	0.00
Products	0.14	Products	0.05
OECD N America	1.74	Other	5.45
Crude	1.74	Crude	4.83
Products	0.00	Products	0.62
OECD Pacific	5.42	Other Europe	0.01
Crude	5.02	Crude	0.00
Products	0.40	Products	0.01
Africa	0.32	SUMED Pipeline	0.63
Crude	0.27	Crude	0.63
Products	0.05	Products	0.00
China	2.14	Unknown	0.02
Crude	2.13	Crude	0.01
Products	0.01	Products	0.00
Latin America	0.14	S Arabia (Red Sea)	0.10
Crude	0.12	Crude	0.09
Products	0.02	Products	0.02
Grand Total	16.85		
Total Crude	15.54		
Total Products	1.31		
Source: Lloyds Marine Intelligence			

FIGURE 2.5

National dependence on oil transiting through the Straits of Hormuz.

---

trade throughout the world and as intermodal connections to other transportation systems.

Ports as Targets

As the connection between land transportation and maritime trade routes, seaports have historically been identified as strategically important targets. The ability to deprive legitimate users of control or use of a port through seizure or damage has been identified as a key potential threat to the integrity of maritime and intermodal supply chains. Ports form single points of failure for SLOCs, especially where they have unique characteristics or geographic locations that make them difficult or impossible to replace. Ports also serve to channel cargo into predictable and known storage areas, such as storage tanks or container yards. This allows potential attackers or criminals to readily locate potential cargo for theft or damage and to focus on the key operations that could be damaged in order to put the port out of operation if desired. While at sea, ships carrying cargo can, in many cases, be diverted, change course to avoid threats, or delay their arrival in a port if necessary, thus making their location and movements less predictable. However, once in port, ships and their cargoes can be fixed in time and space by potential adversaries. Ports, however, are always fixed in time and space. This makes them a fundamentally attractive target for criminals and terrorists by removing any uncertainty regarding the location of the target.

FIGURE 2.6

Port of Rotterdam.

FIXED IN TIME AND SPACE

This term is used in several different disciplines. It is used in geometry, political science, and military strategy and thinking. Its basic meaning in the context of this subject is the idea that, unlike other elements in the maritime transportation system, ports are permanently placed fixtures that cannot and are not designed to move and/or are therefore limited to one predictable location, and their operation is typically on a daily or yearly basis.

Ports also serve as a focal point for other transportation infrastructure. Freight rail systems often have a terminus in large ports, and regional and interstate highway systems are often designed around large port complexes in order to facilitate the movement of freight and cargo. This means that ports often form the primary freight transportation node in many metropolitan areas. As a result, it is easier to predict the location of high-value cargoes as they move into or out of a port.

Therefore, due to their relatively high visibility, locations in populated areas, and role as both a hub for marine transportation and cargo as well as a key element in intermodal transportation systems, ports are clearly potential targets for anyone who desires to disrupt commerce or damage a nation's economy. Because targeting of ports for attack or disruption does not serve any obvious criminal objectives, it is primarily in the interest of terrorists or state actors.

Ports as Conduits

Because the purpose of most ports is to receive and distribute cargo and freight, they are also vulnerable for exploitation by criminal groups, terrorists, or state actors importing illicit material into or out of a country, region, or city. Further, ports in most cases serve as a border post for maritime trade including passengers and freight. Therefore, the vulnerability of ports to exploitation has an additional national-security implication.

FIGURE 2.7

CBP officers inspecting a container. *Photo credit: US Coast Guard.*

Approximately 90 percent of global trade is moved by sea. This trade consists of everything from raw materials to automobiles, clothing, and high-end electronics. This high volume creates many challenges for governments and other entities that are responsible for ensuring the safe and secure transportation of cargo around the world. Complicating this issue is the increase in just-in-time delivery that was outlined in Chapter 1. Therefore, customs officials and regulators are faced with the challenge of securely moving huge amounts of cargo as rapidly as possible through ports to their final point of delivery. This has historically created significant vulnerabilities in the security of the cargo as well as the ports through which the cargo transits. The two greatest vulnerabilities faced by ports as conduits are cargo theft and smuggling of illegal material or people in or out of a port.

Cargo Theft

Cargo theft has been a problem for imports for centuries. It can take several forms including pilferage or the hijacking of cargoes that have already been offloaded and are leaving or have left a port. Globally, cargo theft is estimated to cost as much as $30 billion per year. While US cargo theft has likely been reduced as a result of some of the security measures put in place after 9/11, cargo theft remains a significant problem in many parts of the world.

Cargo pilferage is usually the result of activity by persons who work within or report to the transportation system. They may be stevedores, company employees, ship crews, or in some cases government officials. Since cargo pilferage is predominantly a result of an internal conspiracy, it is often difficult to detect or to isolate where, within the supply chain, the theft occurred. These acts of theft may be tied to organized criminal groups that operate on a broader scale.

Another form of cargo theft is the outright theft of full shipments. This has most commonly occurred through the hijacking of trucks onto which containers have been loaded from ships. This is usually better planned than cases of cargo pilferage. The hijacking of cargoes is usually focused on specific types of goods, especially clothing and electronics. In these cases there is usually evidence that

---

FIGURE 2.8

Cargo pilferage. *Photo credit: U.S. Department of Justice.*

FIGURE 2.9

Narcotics in a container. *Photo credit: Royal Canadian Mounted Police, 2012. Reproduced with the permission of the Minister of Public Works and Government Services, 2012.*

someone inside the port is targeting shipments for hijacking outside the port. Also in some cases there is evidence of complicity on the part of some of the truck drivers, who may cooperate with the hijacking by conveniently stopping their truck to "go to lunch" and then finding the truck gone when they come back. While these hijackings or thefts may not occur at the port facility, there is clearly a nexus between the port and the hijacking through the targeting of the cargo by people with knowledge of port operations and cargo shipments.

Until 9/11, there was a tendency among the shipping and maritime community to accept a certain level of cargo theft as a "cost of doing business." Therefore, it is difficult to accurately assess the significance of cargo theft, because many cases were not reported to either law enforcement or the insurance industry. As will be discussed in future sections of this book, because the port security codes and requirements that have been implemented globally since 9/11 are primarily focused on protecting the port as a potential target, they have likely had a minimal effect on preventing cargo theft, especially outside port areas.

Smuggling

As noted previously, due to high volume and requirements for quick delivery of goods and freight, ports remain vulnerable to smuggling. Smuggling may include narcotics, people, weapons, or other illegal material.

Smuggling typically occurs by placing narcotics in the cargo or inside a shipping container on a ship. When the ship arrives in the destination port, the illegal material is either removed from the cargo in the port or from the cargo or container at its destination.

Narcotics and other illegal material can also be smuggled into or through a port by being attached to a ship with a parasitic device. These devices are used to smuggle primarily cocaine, marijuana, and in some cases heroin in watertight containers called parasitic devices that are attached to the exterior hull of a ship below the waterline or inside a rudder compartment. In these cases, the ship crew has no idea that the devices are attached, and when the ship reaches the destination port, the device is retrieved by divers who recover the drugs.

While much energy is expended on combating the smuggling of goods that are inbound, ports are vulnerable to the smuggling of outbound material as well. In the US, outbound smuggling primarily consists of currency, weapons, or technology that is controlled and whose exportation must be approved by the US government. There will be a much fuller discussion of these threats in the next

---

CASE STUDY

In October 2001, Rizik Amid Farid, a Canadian citizen of Egyptian descent, was discovered by Italian authorities in a container on a ship that arrived in Italy. The container he was in had a bed, toilet, and laptop computer. Mr. Farid was an identified person of interest by the FBI and the Royal Canadian Mounted Police due to his potential involvement in terrorism.[3]

FIGURE 2.10

A parasitic device. *Photo credit: US Immigration and Customs Enforcement.*

chapter, along with consideration of the use of maritime smuggling by terrorists.

Ports as Borders

Because ports serve as a conduit for international trade, both legal and illegal, they are important to consider from a border-security perspective. At its simplest, when a ship, regardless of nationality, arrives at a port from an international destination, it is essentially a border crossing between the last port of departure and the port of arrival. For example, if a ship loads cargo in Indonesia and then sails to an American port without stopping at any other port along the way, there is a temporary border between the United States and Indonesia when that ship arrives. This is because security on that ship and the knowledge of any legal or illegal material it may be carrying are only as good as the security in the country from which the ship last sailed. In some ways this concept is the same as that of a truck crossing from Canada to the United States but on a larger and far more complex level. Instead of the truck crossing from one developed, well-regulated nation to another, international maritime shipping involves border crossings that are continually changing and nations at various levels of development and varying levels of maturity of their port-security measures.

Intermodal Connections

Another factor the makes ports vulnerable, both as targets and conduits, is the fact that most of them serve as hubs for intermodal transportation connections. This means that many ports serve as a terminus or major location in highway systems, rail systems, pipelines, other short sea shipping systems (seaborne distribution within a dispersed port area), and sometimes airports.

Almost all cargo arrives at seaports and is then transferred to another mode of transportation. Typically, that mode is either truck or rail. In order to accommodate these other modes of transportation, particularly in a large port with lots of cargo, there are typically sophisticated highway interchanges or nearby railyards specifically designed to support cargo coming into and leaving the port. It is also not unusual to find that the terminus of a petroleum or gas pipeline is at a port. Finally, in some ports, cargo that arrives from international

FIGURE 2.11

Highway interchange near a port.

FIGURE 2.12

Barge in a European port.

locations may be loaded onto smaller ships or barges for distribution within the region by water. This is known as short sea shipping and is most highly developed in Europe, specifically in Rotterdam and other nearby ports.

This chapter has established that ports are extremely complex, particularly vulnerable, and vital to global trade. From a security perspective, ports always need to be assessed from several dimensions: target, conduit, and border. Further, ports are a critical node within the global maritime transportation system because they are fixed, permanent locations that are both more vulnerable and less easily replaced then other elements.

REFERENCES

1. Lipton, Eric, "New York Port Hums Again, With Asian Trade," New York Times, November 22, 2004, p. B5.
2. "World Oil Transit Chokepoints," US Energy Information Administration, http://www.eia.gov/cabs/world_oil_transit_chokepoints/Full.html, Accessed on 28 August 2011.
3. CBC News, "Egyptian stowaway had Canadian passport," http://www.cbc.ca/news/canada/story/2001/10/25/stowaway_farid011025.html, Accessed on 2 September 2011.

PART TWO

Threats to Ports and the Maritime Domain

CHAPTER 3

Threats

INTRODUCTION

The term "threat" is often confused or used interchangeably with other terms such as "risk" or "vulnerability." It is important that the definition of "threat" be clearly understood. A threat is an act or actor that may bring harm or damage to a country, organization, person, or facility. Therefore, the key component of a threat is action or the potential for action. When we assess threats, we are assessing the potential for someone or something to act in a harmful way against whatever we are protecting. Within the context of maritime security, threats will consist of possible harmful or damaging activities carried out by nation-states and their proxies and/or terrorist and criminal groups or individuals not acting on behalf of a nation.

With the fall of the Berlin Wall and the dissolution of the Soviet Union, global geopolitics has become, in many ways, more complicated. During the Cold War there were two major groupings of allied nations that maintained an uneasy balance of power in the world. Nonstate actors such as terrorists and insurgents existed but were typically under the influence of national powers. The primacy of the nation in international security measures also served to provide greater controls over criminal activities. This arrangement led security practitioners to focus on threats from national governments, states, or organizations under their control or influence—their proxies.

FIGURE 3.1

The fall of the Berlin Wall. *Photo credit: Sue Ream.*

Coincident with the end of the Cold War was society's entry into the Information Age. The rapid, and still increasing, development of (and reliance on) information technology and the resulting explosion in electronic communications and commerce have contributed to globalization, which, in turn, has increased the virtual porosity of national borders. This combination has led to the erosion of the ability of national governments to retain full control of some activities within or crossing their borders and has subsequently led to globalization in which interdependence, regardless of national borders, is increasingly the norm.

Over the last two decades, the suite of threats has magnified significantly as terrorist groups, militants, insurgents, and other groups that are not under the control of a nation have increasingly acted unilaterally. In addition to the threats posed by nation-states, terrorists, and other insurgent groups are the age-old threats posed by criminal groups. Crime and criminal activity have been present in ports throughout history. Criminal activity may include smuggling, theft, environmental crimes, the violation of fishing laws and regulations, and the violation of international embargoes.

Further, all of these threats are affected by the level of corruption that exists in ports and in national governments. Therefore, transparency of governance and integrity of security forces are also issues that affect the threat reports on the maritime domain.

THREATS BY STATES

State Actors

Potential state adversaries may consider ports, as part of a nation's critical infrastructure, to be a legitimate target in both a conventional military campaign as well as in asymmetric attacks.

Conventional Military Attacks Against Ports

One of the most well-known examples of one nation's attack on another nation's ports is the Japanese attack on the American naval fleet at Pearl Harbor, Hawaii, on December 7, 1941. This attack, of course, led to the entry of the United States into the Second World War. During World War II, navies from both sides also attacked enemy merchant shipping in order to disrupt its supply chain. In

FIGURE 3.2

The Japanese attack on Pearl Harbor.

these examples, Japan attacked Pearl Harbor in order to inhibit the projection of American naval power into the Pacific, and combatants on both sides recognize the value of targeting merchant shipping due to its role in providing supplies for their respective war efforts.

A more recent example of a nation attacking the port of another nation is the 2008 attack on the Georgian port of Poti by Russian military forces. This attack not only resulted in significant damage to the nascent Georgian navy and coast guard but also led to the brief shutdown of the port to commercial maritime shipping while it was occupied by Russian troops.

An example of a nation-state's potential use of military units to disrupt ports and shipping can be found on the Korean Peninsula. US military planners have assessed that some units of the North Korean Special Purpose Forces have been assigned the mission of attacking, destroying, or capturing South Korean commercial ports in the event of renewed hostilities. The North Korean intent in assigning units to this mission is undoubtedly to prevent the United States and other allied nations from resupplying South Korean and American forces as well as to economically damage South Korea.

Conventional Attacks Against Supply Chains

The land and maritime supply chains that support seaports have also been the target of attacks by conventional military forces. Among the most well-known were the attacks by German naval forces against allied shipping in the Second World War. During the war, the German Kriegsmarine made a concerted effort to disrupt the movement of war materiel and troops between North America, Great Britain, and the Soviet Union, primarily through a concerted campaign to

---

FIGURE 3.3

A sunken Georgian navy ship.

FIGURE 3.4

A merchant ship sinking after being torpedoed. *Photo credit: US National Archives and Records Administration.*

---

attack convoys of merchant ships by submarines. While ultimately unsuccessful, this sustained campaign against vital Sea Lines of Communication posed a direct threat to the allied war effort.

Since the defense against conventional military attacks is the responsibility of the nation's military forces, it will not be considered in detail in this book. However, the protection of ports, the supply chain, and the maritime domain against asymmetric attacks will be addressed.

Asymmetric Attacks

An asymmetric attack is an attack whereby a militarily weaker adversary targets the vulnerable elements of a stronger adversary's infrastructure rather than engaging in direct military conflict. Asymmetric attacks may be carried out by government agents or unconventional military forces using tactics that may include sabotage or terror to damage or inhibit the use of ports and other maritime transportation elements. An asymmetric attack will rely on surprise and will focus on attacking the most vulnerable elements of a target in unexpected ways. These attacks could focus on the supply chain for a port or region, the utilities that provide power and other services to a port, information systems that allow for efficient commerce, or the people who work in the port area. Asymmetric attacks are particularly difficult to identify or defend against for several reasons. It may not be clear which agency or organization should take the lead in developing protection strategies against asymmetric attacks; these attacks may not initially reveal themselves to be intentional acts; and the tactics and techniques that may be used in an asymmetric attack are difficult to predict. Further, it is possible that government capabilities and priorities are organized to respond to potential attackers and attack methods that have been identified and for which plans have been developed. This may lead to a less efficient or effective response by governments against asymmetric attacks.

While there are several examples of asymmetric attacks in recent history, including Al Qaeda's attacks in September 2001, the most enduring illustrative example comes from ancient Rome *(see the Case Study on the following page)*.

The release of oil into the Arabian Gulf by Saddam Hussein's retreating military forces in Kuwait is also arguably an example of asymmetric warfare. Saddam knew that he would not be able to directly confront Coalition forces but understood that the release of large amounts of oil into the Gulf would serve two purposes: to potentially inhibit the ability of allied naval forces to conduct amphibious landings and to deprive Coalition forces of the ability to use the oil to continue operations. While Saddam's efforts were ultimately unsuccessful on all counts, this serves as an example of a state engaging in asymmetric measures to counter a larger opposing force.

Additionally, the Iranian government has publicly threatened to close down the Straits of Hormuz, a strategic maritime chokepoint at the entrance to the Arabian Gulf, if attacked. By closing down the

CASE STUDY

In the mid-1990s the Commandant of the United States Marine Corps, Charles Krulak, recognized that the post-Cold War era was going to require a different approach in military thinking. He was particularly concerned about the increasing viability of asymmetric warfare. In articles and speeches, General Krulak cited the Battle of Teuteborg Forest as an example of what happens to a conventional military force that is unable to adapt to new forms of warfare. In the year 9 CE, a Roman general by the name of Quintillius Varus led Roman Army legions into Germany to quell a rebellion. Varus had previous experience fighting the German tribes and used the same tactics that he had previously used successfully. In this case, however, the German tribes had changed their tactics and drew the Roman legions into an area where they couldn't effectively use their standard large unit tactics and maneuvers. As a result, Varus' superior and better-trained and -equipped army was largely annihilated. Before he was killed, Varus was heard saying, "Not like yesterday, not like yesterday." This neatly captures the problem of asymmetric threats, their inherent unpredictability, and the difficulty of trying to effectively defend against threats that can quickly develop and appear with little or no warning.

Straits, the Iranian government would effectively shut down critical Sea Lines of Communication between the oil-producing Gulf countries and their customers throughout the world. If carried out, this would be an act of asymmetric warfare in which direct military-to-military confrontation was avoided (at least initially) but the energy sources for several economies and countries would be cut off, with the attendant global political and financial impacts.

Finally, the Al Qaeda attack on the USS Cole in October 2000 in Aden, Yemen by terrorists in a small boat loaded with explosives is a more recent example of an asymmetric attack by a militarily weaker adversary against a target that is conventionally powerful.

FIGURE 3.5

Quintillius Varus.

FIGURE 3.6

Saddam Hussein's oil spill.

FIGURE 3.7

USS Cole after being attacked in Yemen.

_____

State Proxies

In addition to using conventional or unconventional military forces, nations may use proxies to carry out attacks against ports and other maritime infrastructure. Proxies are typically organizations or groups that are not formally attached to a national government or government agency. Proxies may include groups from a third country that are under the control or sponsorship of a nation or groups that are allied with a government's position and/or are willing to cooperate with the government in specific areas of mutual interest.

An often-cited example of a state proxy organization is Hezbollah. Hezbollah has close ties with the Iranian government as well as the Syrian regime of Bashar al-Assad, and, while it is often accused of carrying out assignments on behalf of Iran or Syria, Hezbollah also has specific and unique organizational goals. Hezbollah has strong cooperation with Iran and Syria and serves as an agent for their activities but also operates unilaterally in pursuit of its own interests, primarily within Lebanon.

Pakistan's Directorate for Inter-Services Intelligence (ISI) also uses several proxy organizations in the pursuit of its objectives. The attack on Mumbai, India, on November 26, 2011 by elements of Lashkar-e-Taiba, a Pakistani-based terrorist group, was reportedly conducted at the behest and under the direction of the ISI. Further, the Haqqani Network of militants in Afghanistan, which targets Coalition and Afghan government forces, reputedly receives direction and support from ISI.

Proxy Tactics

Proxies may use the same tactics as unconventional military forces or terrorist and criminal groups to accomplish their aims. Hezbollah proves to be a good example of a proxy organization that uses a wide spectrum of tactics in pursuit of its aims. These activities include:

- Conventional military or paramilitary attacks
- Terrorist attacks against nonmilitary targets
- Criminal activity in support of fundraising
- Social support for constituent population
- Participation in political processes

FIGURE 3.8

A raft used by the Mumbai attackers. *Photo credit: © India Today/ZUMA Press.*

Most notably in the 2006 war between Hezbollah and Israel, Hezbollah's military forces engaged in guerilla warfare against Israeli Defense Force units in southern Lebanon. This was coupled with indiscriminate rocket attacks against Israeli civilian targets.

Hezbollah has a long history of terrorist attacks against non-combatants. These include kidnappings, assassinations, and bombings. Hezbollah engages in or facilitates smuggling of contraband in order to supplement the financial support it receives from Iran as well as to cultivate illicit avenues of entry into other countries that could be exploited in preparation for terrorist attacks. Activities include the smuggling of narcotics from areas of production it controls in Lebanon's Beka'a Valley, car theft in Europe and reselling in the Middle East, cigarette smuggling in North America, and numerous illegal activities in areas such as South America's Tri-Border region.

However, Hezbollah also operates schools, hospitals, and other institutions in Shia areas of Lebanon. After the 2006 war with Israel, Hezbollah was proactive in providing reconstruction funds to affected areas and assisted Sunni and Christian Lebanese populations in addition to its Shia constutuency.

Hezbollah's political wing has 12 seats in the Lebanese Parliament and, as of January 2011, was part of Lebanon's ruling coalition, the March 8 Alliance.

As these examples demonstrate, the security threats posed by proxy organizations are not limited to violence but can include criminal activity and political and social activities that serve to strengthen the positions of their sponsors. While Hezbollah is a particularly sophisticated example, other proxies have some, if not all, of these capabilities.

Nonstate Actors

Organizations may engage in temporary or loose alliances with some nation-states but are committed to activities that support the pursuit of their own goals and objectives. Alliances with nations may occur when a nonstate group's objectives coincide with the objectives of a nation. Additionally, in some cases, independent nonstate actors will carry out attacks or other activities in support of nations in return for logistical support.

CASE STUDY

In May 2011, the *New York Times* reported that the Revolutionary Armed Forces of Colombia (FARC) was asked by the government of Venezuela to provide training to Venezuelan militia forces and to carry out assassinations on behalf of the Venezuelan government in return for logistical support in purchasing weapons.[1] This exchange was between equal partners that, while sympathetic to each other's objectives, are not in a formal alliance or proxy relationship. This relationship was one of mutual convenience and benefit by ideologically compatible organizations but was not a relationship where one participant was subservient to the other and was not permanent in nature. Therefore, Venezuelan support for FARC does not portend that FARC will, outside the boundaries of the specific agreement to provide training and operational support in exchange for weaponry, become a proxy of Venezuela. While FARC may support the current Venezuelan government in certain areas, it is not fully beholden to the Chavez regime and should not be assumed to be an instrument of Venezuelan national policy.

In performing threat assessments and analysis, it is critically important to understand the difference between a proxy group that is under the full or partial control of a nation-state and a nonstate actor that may engage in occasional, temporary alliances with nation-states but otherwise retains autonomy of action and separate objectives and goals. Misinterpretation of these relationships can lead to false conclusions regarding terrorist or criminal group interests in exploiting or attacking a particular target or set of targets.

The line between an organization being a proxy for a nation and being a nonstate actor can be blurry and ambiguous. Therefore, it is important to fully understand the relationships of nonstate actors with nations in order to accurately assess the potential threat posed by a group. A more complicated example of this periodic cooperation between independent nonstate actors and nations is found in the murky relationship between Iran and Al Qaeda.

Iran and Al Qaeda should have almost no strategic interests in common except for an antagonism toward the West. Iran is a regional power that is governed by a militant Shiite Muslim regime, while Al Qaeda is a nonstate actor that adheres to a radical Sunni Muslim philosophy. In fact, in several areas, there is evidence that Iran and Al Qaeda have been in direct conflict. This is most evident in Iraq, where Al Qaeda militants have targeted Iranian-supported Shia militias and populations for attack. However, there is also some evidence of limited support for Al Qaeda by the Iranian government in facilitating the movement of Al Qaeda members, money, and other support to and from Afghanistan and Pakistan through Iran.[2]

This is likely based on a calculation by the Iranian government that, despite ideological differences, support for Al Qaeda in destabilizing the efforts by the US and other western nations in Afghanistan is a worthwhile endeavor in order to reduce western influence in the region. Therefore, while a nonstate actor may find it beneficial to engage in specific areas of cooperation with nations that do not share identical goals, limited cooperation does not mean that there is a full alliance. Indeed, despite the limited logistical cooperation between Iran and Al Qaeda, there remain areas of dispute and disagreement. In the September edition of *Inspire* magazine, the English-language magazine published by Al Qaeda in the Arabian Peninsula, Abu Suhail complained about the Iranian regime's insistence in claiming that the terrorist attacks on New York and Washington in 2001 were carried out by the US government. Abu Suhail stated that the Iranian government wasn't truly anti-American and that Iranian President Mahmoud Ahmadinejad's claims of an American conspiracy theory was borne out of jealousy of Al Qaeda's success.

The two primary activities of nonstate-affiliated organizations (nonstate actors) that are of concern to maritime-security practitioners are terrorism and criminal activity. Additionally, the issue of maritime piracy, which is basically a form of criminal activity, is unique because of its occurrence in areas of the high seas that are not subject to any nation's territorial jurisdiction. Further, piracy thrives in areas where the rule of law is weak or nonexistent. Because of the nature of these groups, any of their activities will be asymmetric in nature.

FIGURE 3.9

Inspire's **September 2011 cover.**

Therefore, the characterizations that were applied to state actors and their proxies do not apply.

Terrorism

Nonstate-affiliated or -sponsored groups engage in terrorist activity in order to pursue their agendas. Although there is no internationally accepted definition of terrorism, most definitions identify some common attributes:

- Use of violence or threatened use of violence
- Intended to advance a political, religious or ideological cause
- Influencing or intimidating a government or population

While most definitions of terrorism include the aforementioned elements, there is still significant room for interpretation. The commonly used adage "one man's terrorist is another man's freedom fighter" accurately demonstrates the role of perception in determining and defining terrorism. For example, the Irish Republican Army is labeled as a "proscribed terrorist group" by the British Home Office but has been championed by otherwise law-abiding Irish-Americans including by New York Congressman Peter King. In almost every case, any organization labeled as a terrorist group will have that label challenged by supporters.

This ambiguity in the definition of terrorism is part of the reason why it is difficult for the international community to develop a globally accepted, legally binding definition. Therefore, for the purposes of this book, terrorism will be defined as "the use or threatened use of violence to advance a religious or ideological cause through the intimidation of a population and the targeting of noncombatants." This definition excludes criminal activity for personal or financial gain and conventional military activity between combatant forces.

Criminal Activity

Maritime-security practitioners need to also be cognizant of the threat of criminal activity. This activity may be in support of terrorism or may be for personal or organizational financial gain. Almost any illegal activity that does not meet the definition of terrorism as provided above meets the definition of criminal activity. For the purposes of this book, criminal activity will involve any illegal activity that does not meet the criteria of terrorism as defined previously. Criminal activity can be motivated by financial gain for either a terrorist or criminal group or for an individual. Further, criminal activity can include illegal activities that are not motivated by financial gain but are carried out as a result of mentally or drug-impaired judgment or personal grievances.

Criminal activity includes actions such as smuggling, theft, corruption, trade-regulation violations, and any other illegal activity found in the maritime or port domain. Further, illegal civil disturbances may be included in illegal activity where appropriate.

A nonexhaustive list of examples includes:

- Cargo theft
- Extortion
- Robbery
- Trafficking of people, drugs, stolen goods, weapons, or money
- Hijacking of vessels or vehicles
- Corruption
- Embargo violations
- Customs violations

Cargo theft is incredibly common and lucrative throughout the world. In 2010, it is reported that the loss due to cargo theft in the US was approximately $171 million dollars.[3] While these statistics include all forms of cargo theft, most of the products enter the US through seaports.

Extortion includes illegal activities by criminal organizations in ports and their environs, in which normally law-abiding citizens are forced to provide material, services, or money to organized crimi-

FIGURE 3.10

Seized contraband.

nal groups. Many of the organized criminal groups that can be found in the US or other parts of the world have a presence in seaports. Examples include Mara Salvatrucha, or MS-13, and various outlaw motorcycle gangs.

Robbery in and around seaports is usually a result of the urban and industrial locations of may seaports. Robbery is also occasionally part of the tactics involved in cargo theft, especially cargo hijackings.

Trafficking is one of the most common crimes in seaports and the maritime domain and may include the trafficking or smuggling of persons, money, drugs, weapons, or other contraband or controlled materials.

The maritime trafficking of people involves smuggling people into countries via ships or small boats. Traffickers may either charge the person to be smuggled a cash fee, or the person may be expected to work off the trafficker's fee in labor upon arrival at the destination country. This process is by its nature exploitive and often results in indentured servitude or virtual slavery. Men are often forced to work for almost no money for years, and women and children are often forced into sexual servitude to pay their debts.

The trafficking of drugs and weapons is often associated with organized criminal groups and can, in some cases, be associated with fundraising for terrorism. While most drug trafficking is for financial profit, some is intended to support terrorist groups. For example, in December 2011 a Lebanese man, Ayman "Junior" Joumaa, was indicted in the US for smuggling cocaine and laundering money as part of an intricate plot that involved both raising money for Hezbollah and laundering money for the Mexican "Zeta" drug cartel.

Hijackings of vessels are typically linked to piracy, which is criminal activity. In the 1990s most cases of vessel hijackings occurred in the South China Sea or the Straits of Malacca and were carried out by Indonesian organized crime groups or the Abu Sayaff group in the Philippines. The Indonesian ship hijackings focused on low-value bulk cargos that were hard to trace. After the hijacking, the cargo was sold and the ship was given a false identity and then loaded with more cargo, which was then also sold, with the profits going to the organized criminal group.

While the Abu Sayaff group is a self-professed ally of Al Qaeda, much of its activity is hijacking fishing boats and kidnapping the crews and appears to be designed for fundraising in support of the group's activities. More recently, Somali pirates have engaged in the hijacking of large commercial ships including cargo ships and tankers. The vessels are released after paying multimillion-dollar ransoms.

The level of corruption in ports varies, depending on the country or countries involved. Some countries with ports have a very low level of corruption, which ensures that border and maritime-security measures are carried out with integrity, but others, while ostensibly adhering to international security requirements, cannot be trusted to effectively carry out their security obligations. The level of corruption is important due to the multinational nature of maritime transportation. If a country's security forces do not meet a level of integrity that provides confidence that the country is carrying out its international security obligations, then it is incumbent on the next country to which a ship is traveling to carry out more stringent security measures due to a lack of credibility on the part of security forces in the previous port. Transparency International ranks corruption among

---

FIGURE 3.11

A US Coast Guard drug seizure. *Photo credit: US Coast Guard.*

FIGURE 3.12

Hijacked crew of the M/V Faina. *Photo credit: US Navy.*

nations and issues an annual report that highlights the level of corruption for each country.

For example, according to Transparency International, the United States has a corruption perception level of 7.1 out of 10 (10 being least corrupt and 1 being most corrupt), while Venezuela has a corruption perception level of 1.9.[4] Therefore, due to the level of corruption in Venezuela, the confidence in the ability of the Venezuelan government's integrity in carrying out its port and border-security responsibilities must be questioned, and any ships arriving in the US from Venezuela will likely be assessed as having been in a port with poor or unreliable security.

Embargo violations involve the use of maritime conveyances to illegally trade in banned products or with countries that have had sanctions imposed on them by the international community. For example, a series of sanctions and embargos has been instituted against Iran in response to Iranian efforts to develop nuclear weapons and ballistic-missile technology. These sanctions prohibit trade in certain technologies, oil, and other goods in an effort to curtail Iranian proliferation activities. However, in most cases where embargos are instituted, unscrupulous traders and smugglers will, for a substantial profit, arrange to transfer prohibited cargos.

Customs violations, as opposed to smuggling, involve the misrepresentation of cargos in an effort to minimize the amount of duties that have to be paid. This is often done by claiming cargos to contain different products or be of lower value than they actually are in order to minimize the amount of money that has to be paid.

Piracy

Piracy is a form of criminal activity primarily found in the maritime domain. The United Nations Convention on the Law of the Sea, Article 101, defines piracy as "any illegal acts of violence or detention, or any act of depredation, committed for private ends by the crew or the passengers of a private ship or a private aircraft, and directed on the high seas, against another ship or aircraft, or against persons or property on board such ship or aircraft; against a ship, aircraft, persons, or property in a place outside the jurisdiction of any state; any act of voluntary participation in the operation of a ship or of an aircraft with knowledge of facts making it a pirate ship or aircraft."[5] While the definition of piracy clearly specifies a focus on criminal activity or "private ends," the most unique element of the definition is that the act of piracy can only occur "in a place outside the jurisdiction of any state." Criminal acts of a similar nature that occur within a state's jurisdiction are typically described as "armed robbery at sea" or hijacking. Of course, to the mariners involved, the legal definitions are merely semantics. Therefore, while piracy has gained significant attention over the last several years with attacks in the Indian Ocean, Gulf of Guinea, Straits of Malacca, and South China Sea, the bottom line is that piracy is primarily criminal activity. There are allegations that some pirate activity may be funding terrorist or other activities not purely criminal in nature.

Much of the current discussion of piracy revolves around three particular areas:

- How to address the issues of weakened rule of law ashore that creates environments that are conducive to piracy
- How to coordinate the response to piracy when it occurs on the high seas
- How to protect ships from piracy

While this book will not address the potential tactical solutions to the above issues, the larger issues of strengthening the rule of law both on land and sea as well as the coordination of counterpiracy activities in the high seas demonstrate the unique nature of this criminal activity and its implications for international law and national-security policies. Piracy can be combated through the application of military or maritime law-enforcement power, but this approach provides temporary relief to ships and their crews and does not solve the larger, systemic issues that allow piracy to thrive. Those issues include building governance capacity in the affected regions; ensuring the targeted and effective delivery of international aid in the form of training, equipment, and sustainment; and strengthening the rule of law and providing alternatives to piracy.

Terrorism, State Actors, and Criminal Nexus

In many of the criminal activities listed above, there is a potential or demonstrated nexus with terrorism because criminal activity could be exploited to raise funds or provide logistical support to terrorist activities or the activities of state actors and their proxies. Because of this lack of clear delineation between crime, terrorism, and asymmetric conflict, any maritime-security program or approach must be versatile enough to address both crime and threats to national security.

REFERENCES

1. Romero, Simon. "Venezuela Asked Colombian Rebels to Kill Opposition Figures, Analysis Shows," *The New York Times*, May 10, 2011. http://www.nytimes.com/2011/05/10/world/americas/10venezuela.html?_r=1. Accessed on December 18, 2011.
2. Cooper, Helene. "Treasury Says Iran is Aiding Al Qaeda," The New York Times, July 28, 2011. http://www.nytimes.com/2011/07/29/world/29terror.html. Accessed on December 18, 2011.
3. Stanfill, Joshua. "2010 NICB Identified Cargo Thefts,"National Insurance Crime Bureau, May 2011.
4. Transparency International. "Corruptions Perception Index 2011," http://cpi.transparency.org/cpi2011/results/. Accessed on 18 December 2011.
5. United Nations Convention on the Law of the Sea, 1982. p. 61.

PART THREE

Current Approaches to Maritime and Port Security

CHAPTER 4

Approaches to Security Policy Development

INTRODUCTION

Port-security policy and regulation are influenced by several different issues and interests, which can be at odds with each other. These include the need for commerce to move quickly and efficiently, the political requirement to show aggressive, and proactive approaches to security, an unclear or unrealistic understanding of the related security risks and therefore, an inability or refusal to accurately determine risk tolerance, and varying approaches to evaluating the effectiveness of security measures or programs.

Further, the international community, while adopting the International Ship and Port Security Code and ISO 28000, has divided into two distinct schools of thought. One, led by the European Union, focuses on fully understanding the potential risks posed by participants within the maritime supply chain and ensuring its integrity, while the US approach focuses on application technology to screen and monitor cargo and persons.

POLITICAL CONSIDERATIONS

Since the attacks of September 11, 2001, which drove the intense focus on port and maritime security, there have been associated pressures placed on political leaders throughout the world that have affected the implementation of security measures. These pressures came from a number of sources including commercial interests and domestic political constituencies.

Commercial Interests

While businesses and other commercial interests tend to acknowledge the importance of implementing both port and supply-chain security measures, the overriding concerns of the private sector focus on the costs of implementation of these measures, the increased oversight of government on commercial operations, and the potential delays in the movement of goods and people as a result.

Costs of Implementation

The overarching imperative for commercial entities is to be profitable. Therefore, any expenditures that do not directly enhance profitability are perceived to be cost centers or overhead, which detract from the company's reason to exist. While companies may understand the intrinsic value of the government's role in protecting commerce, national security, and transportation, they often chafe at requirements to implement government-mandated regulations.

In the case of port and maritime security, this is exacerbated by the international nature of the industry and the need for companies to be competitive not only within their countries but also internationally. In the developed world, this concern is increased due to already higher labor rates and a more mature regulatory environment that requires companies to expend funds to comply with social, health, safety, environmental, and security regulations that are not as stringent in other parts of the world. In the United States, the federal government has provided some port-security grant funds to the private sector to help defray the costs of new regulations, but, as will be discussed in a subsequent chapter, the programs have not been particularly effective and appear to be a hodgepodge of grants funding technological acquisitions with little or no funding for maintenance, training, or personnel costs.

Increased Government Oversight

Companies are often skeptical of increased government oversight of their operations and worry about the increased costs of compliance

THE COST OF PORT SECURITY IN THE US

Between 2002 and 2011, the US government has provided almost $2.5 billion dollars through the Port Security Grant Program.[1] Of this amount, some has been provided as full grants, while a portion has required the private sector, state, or local government recipient to agree to a cost-sharing plan. During the same period, there was an estimated total investment of $6.8 billion dollars in maritime security.[2] Therefore, the industry and local or state governments have expended at least $4.3 billion dollars to comply with the post-9/11 security requirements. Of that amount, most is likely to have been expended by the private sector, as state and local governments are also largely dependent on federal funding for homeland-security initiatives.

and potential lost time in ensuring that government audits and inspections are accommodated. This is of particular concern for port operations, where staff may be diverted from their normal responsibilities to accompany or facilitate government regulators.

Potential Delays

Shipping companies and ports use the measure of how quickly cargo moves as a means of developing a competitive advantage over other companies in their industry. This is used in marketing to differentiate themselves from their competitors, especially as delays can add costs throughout the supply chain. Delays can be costly to ships and ground-transportation companies due to the potential for idle time, which costs the operators money. If delays were to become institutionalized and expected, the associated costs of maintaining idle equipment and personnel would likely be passed along to other elements of the supply chain, including consumers. However, if the delays were limited to certain ports, time periods, or countries, they would directly impact the competitiveness of the port or ports affected.

FIGURE 4.1

Cargo trucks backed up at a port.

Therefore, commercial entities have a substantial interest and stake in trying to make port and maritime security measures as unobtrusive and cost-effective as possible while still complying with regulatory requirements.

Domestic Political Constituencies

For legislators and policy makers around the world, the dependence on certain constituencies often influences the development of security legislation. In the case of maritime security, port operations and shipping are generally not well understood by the public. Despite the lack of knowledge there is often a perceived need to appear to be taking aggressive security measures to satisfy political constituencies and the press, regardless of their actual efficiency or effectiveness. This has been labeled by some observers as "security theater," by which they mean that the measures are not necessarily effective but are put in place to mollify elements of the public or interest groups. Much of the criticism of the US Transportation Security Administration's airport passenger screening program is centered on the idea that the program is security theater designed to provide evidence to the public that stringent measures are being implemented, which could be equally effective if less onerous. Often the driver for the implementation of security measures that may be of limited utility but are highly visible may also be the result of a desire on the part of political leadership to create the impression that government is reacting quickly to a threat or potential threat in a way that is easily recognizable by the public.

However, it is important to note that some measures that may be criticized as security theater may in fact have some deterrent value. This will be further explored in additional chapters. Some excellent examples of the political imperative to take action that may not be logical from a practitioner's standpoint can by found in the US experience since 9/11 in two particular areas: container screening and the port-security grant program.

Container Screening

In 2007, the US Congress passed legislation requiring 100 percent of containers bound for the US to be screened in the foreign ports

FIGURE 4.2

TSA screening.

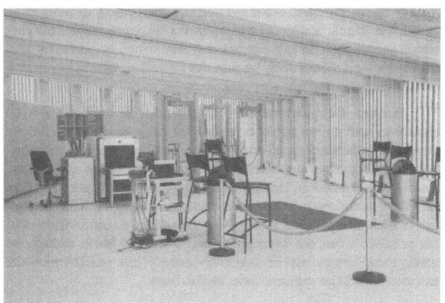

FIGURE 4.3

Container screening. *Photo credit: US Customs and Border Protection.*

of loading by nonintrusive imaging and radiation-detection equipment. While the intent is to prevent radiological or nuclear weapons of mass destruction and other contraband from being introduced to the US, the impracticality of implementing the requirements has prevented compliance. It is predicted that the delays in screening and shipping would significantly impact the US and global supply chains and significantly increase shipping costs.

Further, the mandated screening requirements only apply to containers and do not apply to any other forms of cargo, including general or palletized cargo and bulk cargo. Therefore, the requirement only addresses one form of potential transport of contraband or radiological material. Finally, there is a significant debate regarding the risk of terrorists or criminals smuggling weapons of mass destruction or their components in shipping containers. Counterintuitively, this possibility has not been confirmed as highly likely within the intelligence community or maritime-security risk practitioners for reasons that will be explored later in this book.

Regardless, the requirement to screen 100 percent of all cargo does provide a potentially reassuring message to the public that cargo security is being taken seriously and may be seen as a logical approach to be taken by nonexpert observers.

Port Security Grants

Since their inception in 2002, port-security grants have been provided to private commercial companies, port authorities, and state and local governments to help defray the costs of enhancing security. While the amount of money disbursed has been divided geographically by assessed risk, there has been little effort to ensure that the expenditures have provided tangible benefits to port security. Each year the grant program has been administered, there have been general guidelines that specify how the awarded funds must be spent. This guidance has tended to focus on technologies and equipment until recent years, when training and personnel costs have been permitted, with some limitations. The result has been that equipment, such as x-ray machines and closed-circuit television (CCTV) systems have been purchased but have not been maintained or calibrated due to a lack of sustainment funding. Further, facilities that have acquired these technologies have not been required to demonstrate the effective installation or utilization of the technologies. This is particularly problematic for CCTV systems, as there are fundamental misunderstandings as to how they are intended or able to enhance security. Further, effective CCTV systems require a monitoring capability using smart analytics, since it has been amply demonstrated that human beings are only effective in monitoring a limited number of cameras for less than an hour. Further, as CCTV systems do not respond to incidents, the facility must have an associated response capability or the ability to rapidly notify nearby authorities. These requirements add significantly to the costs of an effective technological security system and have not been covered by the grant programs. Therefore, in many cases, expensive CCTV systems or x-ray machines are installed but are not effectively used to prevent security incidents but rather to provide postincident information to investigators. Further, without maintenance, calibration and updating, these systems can rapidly become degraded or stop functioning altogether.

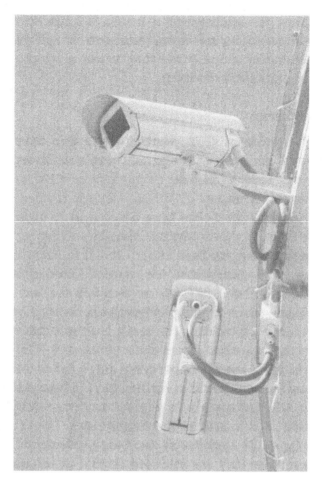

FIGURE 4.4

CCTV system.

The grant program is a result of the pressure placed on political figures and government officials to be perceived as acting aggressively and forcefully on security risks without careful consideration of the effective use of money and the careful development of effective solutions. While the intent of the Department of Homeland Security's port-security grant program and other associated programs is admirable, its execution has been scattershot. The grant program has not required an assessment of the ongoing sustainment requirements for technological acquisitions and associated training and has not been able to clearly provide an indication of the impact that it has had on enhancing port and maritime security. Part of this problem is the result of a lack of an accepted methodology or approach in assessing the impact of security measures.

MEASURING THE EFFECTIVENESS OF SECURITY MEASURES

One of the primary challenges of justifying and validating the effectiveness of security measures, whether in the maritime domain or others, is the ability to measure their effectiveness. While there are established methodologies to measure the effect of security or policing measures on persistent security incidents such as crime, it is particularly challenging to measure effectiveness against less frequent, higher-impact events such as terrorism. While the fundamental concept that needs to be addressed in any consideration of identifying and developing security metrics is deterrence, maritime-security measures, at least in the US, have mostly focused on levels of activity on the part of security forces, with little consideration of the deterrent value of those measures.

Deterrence

Deterrence is the art and science of preventing a person or organization from engaging in unwanted actions that threaten the asset or facility being protected and its ability to continue operations. There are three types of deterrence that are relevant to this problem:

- Punishment
- Denial
- Consequence management

Punishment

Punishment as a deterrent is based on the concept that the imposition of costs or punishment is sufficiently high to make an attack undesirable. Despite the ability to accomplish the desired objective, the potential repercussions are assessed as being high enough to adversely impact the attacker's organizational or personal ability to survive or remain intact. The heightened cost of the potential act serves to deter the commission of the attack or crime. It is important to note that this deterrent type assumes that the organization or person considering threatening activity is concerned with survival. As we will discuss later, the threat of punishment will not be particularly effective against suicide attackers. Further, there is considerable debate among criminologists as to whether an enhanced threat of punishment serves as an effective deterrent against criminal activity.

Denial

Denial minimizes the opportunities available to adversaries. As a deterrent, denial focuses on taking measures to ensure the chances of adversary success are reduced or minimized. If the chances of an attack or criminal activity are reduced sufficiently, then the terrorists or criminals will be dissuaded from focusing on the facility, person, or organization being protected and will move on to targets that offer a greater chance of success. There are several factors that affect the determination of whether denial is a potentially successful approach:

- The uniqueness of the target—if the chance of success in attacking or carrying out criminal activity is limited, are there other targets that the potential attackers could target as alternatives?
- Level of risk that the terrorist or criminal adversary is willing to accept—is the potential adversary willing to accept a high chance of failure or will he or she be deterred?

CASE STUDY

For much of the Cold War, from 1945 until 1991, the United States and the Soviet Union engaged in a strategic nuclear standoff. The primary approach by both sides to ensuring the stability of the international security environment was to guarantee that neither side would be able to maintain its survival if it initiated a nuclear strike. Therefore, the fear of punishment from a retaliatory nuclear strike deterred either superpower from initiating a nuclear war.

FIGURE 4.5

A nuclear-weapon detonation. *Photo credit: US National Archives and Records Administration.*

- Definition of success and failure—how does the adversary define success and do the measures put in place to protect the target reflect those definitions?

Consequence Management

Consequence management involves measures designed to minimize the impact of a successful breach or attack. This approach recognizes that the ability to convince an adversary that an attack or breach is not worth a risk is not successful but that the adversary will not gain a significant advantage by following through with an attack or criminal activity because the target will be able to minimize any impacts and will be able to recover quickly, thereby reducing the importance of any attack. Consequence management includes:

- Rapid response capabilities—security, medical, fire, and utility personnel quickly move to contain the impact of an event and to address the cause as appropriate

FIGURE 4.6

Strong security measures at an industrial facility.

CASE STUDY

In 2002, Iyman Faris, an American of Pakistani origin and Al Qaeda operative, performed surveillance and targeting analysis on the Brooklyn Bridge in New York City, which spans the East River between the boroughs of Manhattan and Brooklyn.

He was tasked with assessing the feasibility of collapsing the bridge by using torches to cut the suspension cables. Due to the heavy security presence that he observed, he reported that the security environment was "too hot" and the plot was aborted.[3] He was arrested in 2003.

- Redundancy—the ability to use alternative or secondary methods to continue vital operations
- Continuity of operations—the ability to maintain key operations throughout an event or to restore those functions rapidly if they have been disrupted

FIGURE 4.7

The Brooklyn Bridge.

CASE STUDY

Since the mid-20th century, Colombia has experienced significant internal conflicts driven primarily by left-wing groups. The most prevalent insurgent or terrorist groups are the left-wing Revolutionary Armed Forces of Colombia (FARC) and the National Liberation Army (ELN). These groups have carried out hundreds of bombings of Colombia's oil pipelines over several decades, with less effective results. The Colombian government and the oil companies involved have taken specific measures to minimize the effects of the bombings as well as to reduce the number of bombings. These measures include the deployment of additional Colombian military troops to protect the pipelines and associated infrastructure as well as measures taken by the oil companies to manage the effects of pipeline disruptions. The oil companies have developed storage facilities in key locations that are well protected and kept full in order to be able to meet delivery and export requirements despite frequent pipeline disruptions. Planning for the continuity of operations, coupled with rapid repair of the pipelines, has led to minimal impact on Colombian oil exports. Despite the serious environmental damage caused by the pipeline bombings and the release of oil into the natural environment, from an economic perspective Colombia has effectively managed the pipeline bombing campaign and reduced it from a potential catastrophe to a persistent nuisance.

It is important to note that while there are three types of deterrence in addressing security issues, the appropriate selection of a deterrence strategy is part of a greater risk-assessment approach that will be discussed in further chapters and will include assessments of the adversaries' goals, objectives, and tactics. Further, the three deterrent measures may be, in many cases, most effective when combined and not implemented in isolation.

Measurement of Activity vs. Effectiveness

Since the attacks of September 11, 2001, in the United States, the methodology to determine the effectiveness of security measures by the Department of Homeland Security has focused on quantifiable metrics instead of qualitative, intelligence-derived assessments of the deterrent value. While not effective in preventing terrorist attacks, the quantitative approaches have strong foundations in both the bureaucratic and political environments and in conventional law-enforcement approaches to crime prevention. However, these methodologies offer fundamentally flawed approaches to the measurement of the effectiveness of security strategies against terrorist threats.

Measurement of Activity

The measurement of activity in security operations takes two distinct approaches, a focus on resources expended and a quantitative approach to crime prevention.

Resources Expended

Among federal "budgeteers" and agency leaders, there is a constant requirement to justify budget and resource requests. This often leads to a need to protect the current budgets from interagency or congressional challenges as well as a desire to expand future budgets. A common approach is to carefully document agency expenditures in order to justify current and future funding levels. This translates into a substitute for actually assessing the effectiveness of the activity but encourages lots of activity of indeterminate effectiveness. An excellent example of this is the legislation passed in 2007 requiring the screening of 100 percent of containers entering the United States. This legislation mandated a high level of activity but did not include any requirement to measure the effectiveness of the legislation. Other common "measures" of mission effectiveness include:

- Personnel hours expended
- Personnel trained
- Patrol numbers and frequency
- Boat hours expended
- Boardings conducted
- Units deployed

The measure of activity is useful for those in government, as it is a simple way to document the amount, especially if tied to plans that call for increased activity and resource expenditures as a means of deterring potential terrorist attacks. The problem with this approach is that there is no analytic justification that ties the increased activity to deterrence against terrorism. As will be shown next, increased activity may be useful in preventing criminal activity, but agencies have not made a strong case that increased activity prevents terrorism by itself.

Measurement of Criminal Activity

The second quantitative method of measuring security-related activity is the measurement of criminal activity. While this model may be useful for assessing the effectiveness of crime-prevention measures and, to an extent, border security, it is less effective in measuring the effectiveness of security against terrorist attacks on civilian infrastructure. There are two widely known tools for measuring crime in the United States.

In the United States, crime figures are measured by the federal government through the collection of data from law-enforcement agencies throughout the country, which are then reported through the Uniform Crime Reporting (UCR) Program. While the UCR Program is very useful in determining detailed crime rates and trends, it does not provide any analysis of the reasons why crime rates may be rising or falling. Therefore, any analysis of the effectiveness of anticrime measures must be extrapolated from the data and is not directly correlated to the Program.

Many municipal police departments use CompStat, an anti-crime-management approach that couples statistical analysis of criminal activity with the development and implementation of strategies to combat areas of high crime. It also incorporates a management approach that makes police managers accountable for the crime rates in their areas of responsibility and provides them with the ability to make decisions to combat crime rates. Developed in the 1990s by the New York City Transit Police Department's Jack Maple and later adopted by the NYPD under Commissioner William Bratton, it has been further adopted by many American police departments and some departments outside the United States. While generally perceived as effective, it has received criticism for pressuring police commanders to underreport crimes in their areas of responsibility, which, if true, harms the integrity of the data. There has also been speculation that crime-rate reductions that were attributed to Comp-Stat in the 1990s may be the result of other factors, including an influx of new, better-educated police officers and the end of the crack epidemic, for example.

Unlike the UCR Program, CompStat is able to demonstrate a cause-and-effect relationship between crime rates and anticrime strategies. Therefore, under CompStat, a degree of effectiveness can be measured.

The reason that a statistical approach to maritime homeland security is flawed, especially when measuring activity, is that statistical models that are geared towards budget issues (i.e., activity

FIGURE 4.8

Police officers on patrol.

reporting) or law enforcement (crime reporting) require significant amounts of data. Therefore, these models, while potentially effective in maritime homeland-security-related missions such as border security, narcotics, and illegal immigration, do not make room for the "black swan" event that is a primary concern of maritime homeland-security practitioners. The "black swan" theory is derived from the book of the same name by Nassim Nicholas Taleb, in which the theory is presented that in the aftermath of seemingly unpredictable high-impact, low frequency events there are indicators that, had they been identified and analyzed, would have made the event detectable prior to its occurrence.

FIGURE 4.9

A black swan.

The prime example of a black-swan event is the attacks of September 11, 2001. While the attacks were unexpected, there were indicators that Al Qaeda was interested in using aircraft against Western and US interests, such as the foiled Bojinka Plot to blow up 12 aircraft over the Pacific Ocean in 1995, coupled with local FBI reporting of suspicious activity by Al Qaeda-linked flight students in Arizona in July 2001. Due to the fact that there were no large amounts of data allowing for the development of patterns of activity CompStat-style, this high-impact, low-frequency event and its indicators would have been undetectable using conventional crime and activity reporting methods.

How to Measure Effectiveness

One of the major questions often raised is why there have been no other black swans in the U.S. since 9/11. While there are many possible considerations, there are no clear answers since the deterrent effect of the security measures instituted since September 2001 are unknown. Possible explanations that have been posited include the following, which may be considered separately or together:

- The effect of expeditionary attacks against Al Qaeda in other parts of the world
- Law-enforcement and intelligence efforts to disrupt logistics and communications
- Al Qaeda's instructions to followers to focus on the "near enemy," Western interests in the Middle East region
- Diffusion of Al Qaeda from a tightly controlled organization into a series of loosely affiliated groups
- The deterrent effect of security measures

We will see in future chapters that one of the primary ways to measure deterrence is in the combination of a CompStat-type approach with an intelligence-focused assessment of the adversary's perception of the effectiveness of security measures. In short, where possible, statistical information compared against security measures should be used, but we also need to assess the enemy through intelligence collection and analysis in order to be able to measure the effectiveness of security against the potential black-swan attack.

Why Don't We Do This Already?

Pockets of government may be making these assessments, but, due to information security sensitivities, the results are not released to the public or the front-line maritime security practitioners who need to plan the daily and weekly protection missions. This forces the practitioners to measure what they can, which is activity and the effects of security measures on lower-level criminal and suspicious activity. These less effective measures are reinforced by agency leaders and the political establishments, who are consumed by budgetary issues and protecting their agencies and interests from bureaucratic attack.

Further, there appears to be a tendency to try to measure the effectiveness of security using the same types of metrics used by safety professionals. Prior to the attacks of September 11, 2001, there were relatively few maritime-security professionals. Certainly in the United States, maritime and port security was perceived as a niche, somewhat marginal profession with little relevance in the post-Cold War geopolitical environment. Therefore, out of necessity, after 9/11, many maritime professionals with safety backgrounds found themselves thrust, either voluntarily or involuntarily, into security positions they were not particularly well prepared for or knowledgeable about. Many of these professionals adopted an approach that "safety and security are two sides of the same coin," meaning that the approaches and techniques that they used for marine safety could be applied to security challenges as well. This led to an acceptance of measures of effectiveness that discounted the lethal threat posed by thinking, cognizant, adaptable enemies bent on intentionally harming people and infrastructure and instead took an approach similar to preventing unintentional accidents. Therefore, measures of effectiveness focused on statistically measurable safety issues were applied erroneously to security problems. Fundamentally, safety and security are not "two sides of the same coin." While there are common response options and protocols that may be combined to realize some efficiencies, the measurements of effectiveness are inherently different.

Another reason is that it is hard and there is inherent ambiguity in dealing with an adaptive, intelligent enemy. Immediately after the 9/11 attacks, the key homeland-security agencies had to quickly shift their missions to a new threat. FBI agents who were primarily criminal investigators or counterintelligence officers had to learn about intelligence collection and analysis as well as a new enemy. Customs inspectors had to shift from an antismuggling and revenue-collection focus to one of literal border security, and Coast Guard members had to shift from their previously normal functions in law enforcement, search and rescue, and maritime safety to reinvigorate their traditional but neglected port and maritime security mission. Therefore, these agencies and their personnel had to learn to implement homeland-security measures at the same time they were performing their missions during a period of grave crisis. This has been described as being akin to changing a tire on a car while it's traveling down a highway. In order to train and deploy large numbers of inexperienced personnel, the normal solution was to teach the "science" of maritime security and measuring the effectiveness of security in traditional terms that are accepted by the political establishment, which became a default position. Teaching the "art" of maritime security, allowing for greater ambiguity and a focus on qualitative assessments such as intelligence reporting of enemy perceptions of intelligence, has apparently been a challenge for relatively inexperienced practitioners and is likely to be viewed with significant skepticism by Washington's all-powerful "budgeteers."

The Maritime Context of Assessing Deterrence

Since there is no widely available information regarding the effectiveness of the maritime-security measures taken to date, it is largely impossible to measure the deterrence value of what bureaucrats call Ports, Waterways, and Coastal Security (PWCS) missions. This, in turn, makes it challenging for Coast Guard operational personnel to assess priorities for deterrence-based missions. These missions include positive control boardings, vessel escorts, safety and security zone enforcement, and security boardings. Further, the associated security regulations (the Maritime Transportation Security Act and the International Ship and Port Security Act) are designed to enhance security at waterside facilities and commercial vessels and are inherently deterrent regulatory regimes. However, the degree of effectiveness of the PWCS mission and associated regulations has not been accurately determined. General information provided through the intelligence community has established the value of deterrence on the terrorist attack cycle, particularly during the surveillance and planning phases. According to intelligence community assessments, Al Qaeda and its allies are most likely deterred by security measures that reduce their chances for success. Punishment is not a credible deterrent for suicide bombers, but the fear of failure is assessed as being credible. Similarly, criminal groups attempting to use the ports as conduits for smuggling contraband and/or people are likely to be deterred by a fear of punishment as well as (secondarily) a fear of failure. However, there appears to have been no in-depth assessment regarding how effective certain tactics have been or which elements of maritime infrastructure are best protected by which methods.

In future sections, a proposed model for the establishment of an effective deterrence assessment program will be introduced. It will show that deterrence in security operations or policies can be most effectively assessed through periodic intelligence assessments regarding our adversaries' (whether terrorist or criminal) *perception* of security and the deterrence value our adversaries give to various regulatory and PWCS measures. Coupled with analytically rigorous and informed "red cell" analysis from an adversary's perspective, using the threat information provided, this approach would be useful in then identifying vulnerabilities and enhancing those security measures deemed to be most effective.

LACK OF A RISK APPROACH

The current state of maritime security, and homeland security writ large, is further burdened by the lack of a true risk-based approach in some areas. A real, risk-based approach to security forces practitioners or owners and operators of affected infrastructure to make difficult decisions regarding which elements, facilities, processes, or people to focus their security measures on. The requirement to put priorities on both resources and elements of the maritime domain is, in itself, a risky proposition and forces decision-makers to commit to a course of action that necessarily favors some elements of the maritime domain over others. Coupled with the inherent difficulty in making risk-based judgments is the challenge of truly understanding security risk. Understanding risk is complicated by several factors:

- A lack of understanding of security risk components
- A lack of a process to determine risk tolerances (and often a lack of understanding of what risk tolerance is)

- A tendency towards risk aversion
- A focus on risk mitigation instead of risk treatment
- The lack of political will among leaders to make truly risk-based decisions
- A lack of recognition of critical nodes in the maritime domain
- Overquantifying security risk
- A tendency to use the rubric of all-hazard risk to simplify an inherently complicated process
- A propensity to minimize the element of threat in performing security risk assessments

What is Risk?

There are numerous definitions of risk. Three of the most common are definitions from the United States, Australia, and the United Kingdom are as follows:

- Risk is defined by the US Department of Homeland Security as the "potential for an unwanted outcome resulting from an incident, event, or occurrence, as determined by its likelihood and the associated consequences."[4]
- According to the Australian/New Zealand Risk Management Standard 4360, risk is defined as "the chance of something happening that will have an impact on objectives."[5]
- According to the British Standards Institute, risk is "something that might happen and its effect(s) on the achievement of objectives."[6]

Of particular note is that the Australian and British definitions treat risk as neutral while the U.S. approach defines risk in a negative context. This different approach to defining risk is likely at the heart of the response to and to the treatment of risk, where the US focus appears to be on mitigating risk as much as possible, while the UK and Australian focus is to approach risk as something that needs to be accepted or treated, with mitigation being one of several options.

The insurance industry, due to its role in insuring risk, has done considerable work in the identification and classification of risk. The industry has categorized risk into four types:

- Dynamic risk
- Pure risk
- Fundamental risk
- Particular risk

Dynamic Risk

Also known as "speculative risk," this type can lead to either profit or loss and is characterized by intentionally acting despite the inherent uncertainty with the desire to realize profit but an understanding of the possibility of realizing loss. Businesses entering new markets, gamblers, entrepreneurs marketing new products or solutions, and criminals are examples of persons or organizations engaging in dynamic risk.

Of relevance to maritime-security practitioners is the understanding that adversaries who are targeting ports and shipping are engaging in dynamic risk. This is important when developing security approaches to treat the risk posed by the potential adversaries. For criminals, an increased likelihood of capture, or failure in the case of terrorists, will affect the likelihood of an attempt against a target that has factored a dynamic-risk approach into its risk assessment. Therefore, understanding an adversary's approach to dynamic risk can inform the types of deterrent measures that need to be put in place.

However, since the current definition of risk in the US Department of Homeland Security does not allow for an understanding of dynamic risk from a policy perspective, as risk by definition can only be negative, the factoring in of such an understanding to enhance deterrence by decreasing the adversary's chance for profit (success) may be problematic for practitioners following the US approach.

Pure Risk

Also known as "static risk," this type can be either negative or neutral. This approach to risk is consistent with the US Department of Homeland Security approach to risk. Pure risks include situations that are typically insurable, such as catastrophic medical bills, automobile accidents, or home and property loss. In all these examples, the best outcome is neutral (the insurance is not necessary but available), while the worst is a need to address the pure risk as a negative event.

This is relevant to maritime security because it encourages the holder of the risk, such as a shipowner, to ensure sufficient coverage to cover the possibility of loss or damage.

Fundamental Risk

Fundamental risks are defined by their scope. They may be either dynamic or pure risks but will affect an entire organization, society, or nation. A relevant example would be a risk to the integrity of the international supply chain.

Particular Risk

Like fundamental risk, a particular risk can be either dynamic or pure. However, particular risk is focused on a specific individual, set of individuals, or facility, such as a specific container terminal or fuel storage depot.

The differentiation between fundamental and particular risks is important, as the approach and role of government are likely to be different

Components of Security Risk

Security risk is generally accepted to consist of three components: threat, vulnerability, and consequence. The term "likelihood" is also used in assessing the relationship between the threat and the vulnerability of the potential target. These may be represented somewhat differently in some methodologies, but the general approach is consistent.

Threat

Threat usually defines the potential source of an action or activity that is being protected against. Some definitions, including those by the US Department of Homeland Security, include natural disasters and other catastrophes as threats. This broader definition, while simpler to use, weakens the analytical rigor needed to focus on security threats by including events that are not controlled by humans who are actively trying to circumvent protective measures. Therefore, the assessment process is given less emphasis, as there is a broader swath of potential crises that cannot be prevented. This lowered emphasis on accurately assessing security threats by mixing them with safety threats has led to a weakening of the threat component in "all hazard" risk assessments and therefore, has drawn focus away from effective and preventive security-risk treatments.

For the purposes of this work, the definition of threat will focus on the potential "bad actors" that knowingly and intentionally target the maritime operating environment for harmful exploitation. All other potential impact events or sources will be termed as "hazards." This is important because in later chapters, we will focus on threat assessments, which are performed in a completely different way than hazard assessments.

Vulnerability

Vulnerability is the ascertaining of where protective or procedural weaknesses exist that may make it easier for potential adversaries, the threat, to exploit or damage the entity, person, or facility that is being protected. Vulnerability can include elements such as fences, walls, or access-control systems as well as response procedures, security-guard procedures, threat-detection capabilities, and the ability to withstand and recover quickly from an attack or intrusion.

Consequence

Consequence is the potential result of a successful attack or intrusion by an adversary due to the exploitation of one or more vulnerabilities. Consequences may include the disruption of operations, the loss of life, resources, and assets, the loss of services, and damage to property.

A simple yet illustrative example of risk might be the case of thief who breaks into a facility to steel copper wire because the perimeter fence was in disrepair. The thief is the threat, the poor condition of the perimeter fence was a vulnerability, and the consequence is that the copper wire is stolen, a financial loss as well as its no longer being available for use. The risk is that the facility might be broken into and material stolen.

Risk Management

While risk has varying definitions, the management of risk follows a generally accepted pattern that includes the following components, which are derived from several international standards including *ISO 31000:2009—Risk Management. Principles and Guidelines on Implementation.*

- Establish the context in which the risk is being assessed
- Identify the risks that the organization may face
- Analyze the risks
- Evaluate the risks
- Treat the risks

There are two further critical elements of the risk management process:

- Communication and consultation with all relevant internal stakeholders to ensure the support of senior management and participation of relevant personnel as well as the determination of risk tolerances
- Monitoring and review of the process to ensure assessments and recommendations remain relevant in light of any changes in the context or risks

These will be covered in detail in future sections of this book as recommended approaches to risk management are explored.

The Weaknesses of Current Risk Management Approaches

Currently, many security practitioners and policymakers are ostensibly advocates of a risk-based, resilient approach to security, but the execution of maritime security on national levels and in the international realm has not so far demonstrated a fully risk-based approach. As noted previously, the main issues that tend to confound the effective implementation of risk management include:

- A lack of understanding of security-risk components
- A lack of a process to determine risk tolerances (and often a lack of understanding what risk tolerance is)

- A tendency towards risk aversion
- A focus on risk mitigation instead of risk treatment
- The lack of political will among leaders to make truly risk-based decisions
- A lack of recognition of critical nodes in the maritime domain
- Overquantifying security risk
- A tendency to use the rubric of all-hazard risk to simplify an inherently complicated process
- A propensity to minimize the element of threat in performing security risk assessments

In the following paragraphs, these issues will be explained in further detail.

Lack of Understanding of Security Risk Components

The most common misunderstanding is the tendency for some practitioners and policy-makers to interchangeably use the terms "risk" and "threat," which leads to confusion regarding the actual issue being considered. This is exacerbated when "threat" is used to include nonhuman actors that are incapable of acting with intent, such as earthquakes and floods.

Lack of a Process to Determine Risk Tolerances

Most risk-assessment methodologies either do not address, or address in a very cursory manner the issue of identifying the risk tolerance of the subject of the risk-management exercise. If it is accepted that risk management is really about managing risk, not just mitigating or reducing risk, the determination of risk tolerance is of vital importance. This process needs to be included in risk assessments so the appropriate treatment measures can be developed to meet the requirements of the protected entity.

Risk tolerance is the amount of risk that can be accepted by a person or entity without the requirement to treat the risk. An established process to include this in risk-management methodologies or approaches will serve to more effectively identify the critical elements of an organization or operation and will likely prevent tendencies to risk aversion.

Tendency Towards Risk Aversion or Avoidance

The lack of risk-tolerance activities in risk assessments reinforces the tendencies on the part of policy-makers and other key stakeholders to not engage in the rigorous and difficult work of determining which risks cannot be or don't need to be eliminated. Part of this tendency to use risk management to avoid risk is a result of a lack of in-depth understanding of risk management as well as the US definition of risk as an unwanted outcome.

Focus on Risk Mitigation (Reduction) instead of Risk Treatment

In those cases where decision-makers are risk-averse, the default position is often one of risk mitigation as the only option. This approach does not allow for a more nuanced approach to risk that would be possible if all of the elements of risk treatment were available. Some of these include:

- Risk retention
- Risk reduction
- Risk avoidance
- Risk transfer

It takes a great deal of confidence in the ability to assess risk as well as political courage to articulate a risk-management approach that includes elements of risk retention and transfer as well as incident- or target-specific risk-reduction measures within specific geographic or operational contexts.

Lack of Recognition of Critical Nodes in the Maritime Domain

In some cases, decision-makers and policy-makers do not have a full understanding of what the critical elements or nodes of the maritime domain are within their areas of responsibility. This is especially true where national-level policy-makers focus on geographical regions or industries in which they are not particularly familiar and about which they do not get sufficient advice from local or industry experts. The lack of local or industry perspective can lead to a misjudgment of the critical elements.

CASE STUDY

In the United States, there are two large ferry systems that carry cars between Long Island, NY, and Connecticut. Some of the ferries can carry up to 100 cars and 1000 passengers. Long Island has approximately seven million inhabitants living in two boroughs of New York City (Brooklyn and Queens) and two suburban counties (Nassau and Suffolk). On September 11, 2001, the bridges and tunnels that connect Long Island to Manhattan, Staten Island, and the Bronx were shut down to all nonessential traffic. As a result, the only way on or off Long Island for seven million people for over 24 hours was via the two ferry systems transiting Long Island Sound. However, because the ferry system straddles two states, the jurisdictions of two FBI field offices, and two Federal Regions, the importance of the ferry system as a key resupply and evacuation asset for Long Island has been overlooked since 9/11, and security measures and resilience investments in the ferry systems have arguably been underfunded.

This is likely due to the lack of any single entity to advocate for the ferries and Long Island as well as the lack of awareness on the part of national policy makers to the significance of the ferry systems due to the fact they have no advocate to make the case among competing national infrastructure elements that do have single and unified advocates.

Overquantifying Security Risk

One approach to security risk assessment is to use mathematical calculations to determine risks to particular assets or areas. This method, rooted in operations research principles, has been incorporated in several US Homeland Security assessment tools. While this methodology is potentially useful in assessing the risks posed by hazards such as accidents or natural disasters, it is flawed when assessing the risks involving sentient actors who are capable of quickly adjusting behaviors based on responses to changing security environments. In some

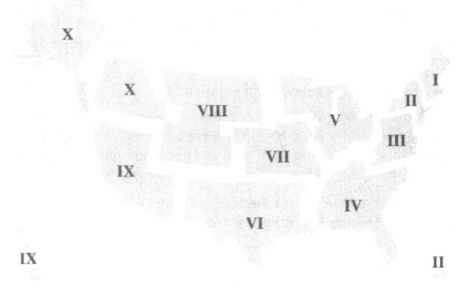

FIGURE 4.10

Federal regions in the US Northeast.

FIGURE 4.11

Long Island Sound Ferry.

models, proponents of the quantitative approach even advocate the ability to use mathematical formulas to "buy down" risk, which is an apparent reference to the focus on mitigating or reducing rather than treating risk.

Tendency to Use the Rubric of All-hazard Risk

As discussed previously in this chapter, assessment approaches that include nonsecurity hazards tend to serve as attempts to simplify an inherently complicated process and, as a result, minimize the threat element in risk assessment, which then skews the assessment's results.

A Propensity to Minimize the Element of Threat in Performing Security Risk Assessments

Sometimes the lack of attention and rigor towards threat assessments results in a tendency towards threat advocacy and the improper use of "red cells." Red cells are used to develop an enemy's or adversary's approach to potential attacks or exploitation of a target. In some cases, they have included Hollywood screenwriters as well as former special operators who are fully capable of planning attacks against targets but may not be able to look at the potential targets through the prism of a terrorist or a street-gang member. In the intelligence community, this is called "mirror-imaging," a situation where threats are assessed by individuals approaching the problem by looking at it through there own world view vs. that of their adversary. A retired commando can provide the best information on how to attack a port and identify vulnerabilities, but it is unlikely that the same special operator could provide meaningful insight as to what the potential targets would be for an Al Qaeda operative or how Al Qaeda would carry out the attack. That is the work of intelligence analysts dedicated to following Al Qaeda.

Threat advocacy is a situation where, in the absence of a fully justified threat assessment, certain individuals will develop "what if" scenarios that are not fully assessed and may be focused on the desire to justify certain preordained views or desired security outcomes. Brian Jenkins, an internationally recognized terrorism authority at Rand, argues that much of the national debate in the US is centered around threat advocacy, rather than threat assessment, in which elements of the national-security community may have concerns that do not correspond with the adversary's actual strategic and tactical goals. This, coupled with the difficulty in conducting threat analysis, has led to the acceptance of vulnerability-based security measures in which vulnerabilities are identified, a possible attack vector is identified and a worst-case scenario is postulated. Jenkins argues that these activities are appropriate for consequence planning but do not constitute a threat assessment.

Part of the tendency in some methodologies to minimize the threat component is rooted in the fact that threat information is often treated as classified or sensitive and therefore is perceived to be a mystery to many security practitioners. In future chapters, a methodology for carrying out threat assessments will be introduced.

REFERENCES

1. "Port Security Grant Program: Risk Model, Grant Management, and Effectiveness Measures Could Be Strengthened," Government Accountability Office. November 2011, p. 2. http://www.gao.gov/assets/590/587142.pdf. Accessed 6 January, 2011.
2. "Port Security Grant Program (PSGP)," US Maritime Administration. http://www.marad.dot.gov/ports_landing_page/infra_dev_congestion_mitigation/intermodal_transport_networks/intermod_trans_net_port_sec/PSGP.htm. Accessed 6 January 2011.
3. Statement of Facts, United States of America vs. Iyman Faris, US District Court for the Eastern District of Virginia.
4. Risk Steering Committee, *DHS Risk Lexicon: 2010 Edition*. September 2010, p. 27.
5. Council of Standards Australia and Council of Standards New Zealand, *Risk Management: AS/NZ 4360:2004*. Section 1.3.13.
6. British Standards Institute, *Business Continuity Management—Part 1: Code of Practice. BS 25999-1:2006*, p. 5.

CHAPTER 5

A Critique of Current Maritime Security Measures and Approaches

INTRODUCTION

Since the watershed of September 11, 2001, maritime-security approaches have included the development of national and international strategies, the implementation of national and international security regulations, and the development of capacity and capability among governmental agencies as well as the private sector. The effectiveness of maritime-security approaches is open for debate.

Among the continuing issues surrounding the evolution of maritime security are the roles and responsibilities of government agencies and the private sector, especially in identifying how maritime-security measures should be funded. There are cogent arguments that support both a government-focused funding approach as well as one that the maritime industry and its components should bear the costs of security. Most common, however, is a recommended approach that balances government expenditures on missions and measures that are its responsibilities with the obligations of an industry to protect itself.

While there have been no catastrophic terrorist attacks involving maritime conveyances since 9/11, there is still an epidemic of

FIGURE 5.1

The evacuation of lower Manhattan on September 11, 2001. *Photo credit: US Coast Guard.*

---

illegal activity on ships and at ports. Further, while the focus on the security and resilience of the intermodal supply chain is shifting and expanding, it remains in its early stages, with competing international approaches but no accepted global standard.

Finally, the rise of piracy in the Indian Ocean and other parts of the world highlights the connections between security on land and the maritime security of adjacent waterways as well as the need for international agreements for security on the high seas or in areas where the rule of law is weak.

Regulations and Their Limits

Regulations, by their nature, are designed to establish a required baseline of performance for the industry or activity being regulated. In the case of regulations that apply to a global industry such as maritime ports and shipping, such as the International Ship and Port Security (ISPS) Code, regulations are typically written in such a way as to be implemented by a wide variety of facilities or vessels. This approach is written towards "the lowest common denominator," or the asset that will find it most difficult to comply with the implementation of standards due to either resource availability or overall capability. Supply-chain security has, due to its nonmandatory nature, a much broader suite of programs and differing approaches to implementation.

The ISPS Code

The ISPS Code is a reasonably effective initial step in establishing low-level baseline security in global shipping. Of particular note is that the ISPS Code is designed to focus on the desired security outcomes without being overly prescriptive in the manner in which the outcomes are realized. This is particularly effective because of the drastic differences in the sizes, technological development, and resources available to ports, administrations, and shipping companies around the world. Further, while the Code requires national governments to oversee the implementation of the Code, it does not prescribe the detailed roles of government and the private sector since those roles will vary from port to port or nation to nation.

FIGURE 5.2

An ISPS audit. *Photo credit: US Coast Guard photograph by Petty Officer 3rd Class Andrew Kendrick.*

However, the ISPS Code has several inherent weaknesses, including:

- The ISPS Code focuses primarily on external threats and does not address supply-chain security issues as a primary concern. The Code's focus on external threats creates a vulnerability regarding the potential for internal conspirators to initiate criminal activity or terrorist acts from inside the fence line of an ISPS-compliant facility. The ISPS Code addresses security in the context of the port or ship as a potential target vs. a conduit for illicit material or people. While there are other supply-chain security initiatives, they are not in all cases tied to the ISPS Code. An exception is the release of ISO 28000 by the International Organization of Standards. Through its subsidiary standard, ISO 20858, ISO 28000 makes the ISPS Code part of a continuous-improvement cycle, following the plan-do-check-act cycle common to internationally recognized quality-management systems. However, there are no national or international requirements to implement ISO 28000.

- The ISPS Code relies on self-regulation. While the ISPS Code required national governments to oversee the implementation of the Code and to approve security plans and training, facilities, ships, and companies are expected to identify their own critical operations and to carry out their own threat and vulnerability assessments. Most commercial operations do not have staff with the skills to carry out these assessments and may be inclined to minimize critical operations in order to reduce the amount of regulatory burden. In many cases, government regulators depend on the judgment of the entities being regulated to determine the level of compliance required. Further, regulators tend to periodically audit the implementation and rely on the regulated entities to report changes in operation or security incidents.

- There is no public information regarding how potential adversaries rate the effectiveness of the ISPS Code. While there may have been an analysis of the effectiveness of the Code by security or intelligence agencies, it has definitely not been shared with corporate security officers or others responsible for the implementation of the Code. Knowing the perception of adversaries, whether terrorist or criminal, would be valuable in both strategic and policy decisions regarding any potential changes that should be made to the Code or its implementation, as well as enhancing the targeted expenditure of government funds for maximum effect. Further, providing feedback to the private-sector security officers and managers responsible for the implementation of the Code would both allow for changes in implementation as well as validate the costs associated with compliance.

- The ISPS Code, written to promote self-regulation, has an inherent assumption that all parties involved in day-to-day port business or operations are legitimate actors. The Code does not provide detailed requirements for the conduct of background checks for employees or authorizations for visitors, thereby creating a significant variance in access to port areas among countries. As a result, several countries, including the United States and Australia, have supplemented the requirements of the Code by creating maritime or port credentials for private-sector persons to enter port areas. The issuance of these credentials is predicated on a background check and threat assessment by appropriate government agencies.

- Regulations compel compliance but do not necessarily effectively build security into daily operations or the organizational/corporate culture. The ISPS Code has been adopted by most countries in the world that have ports or register shipping fleets. In most cases, the ISPS Code is adopted outright and is made regulatory through implementing legislation. Some countries have developed national port-security legislation that exceeds the requirements of the Code. However, in both cases, the Code is required by regulation and compels certain elements of the maritime industry to comply with requirements that can be both costly and inefficient. This approach, which essentially places an externally required set of conditions on ports and ship fleet operators, does not encourage the development of a "security culture" that is interwoven in the daily operations and leadership culture of organizations subject to the Code. Further, there is typically a lack of any positive incentives to participating beyond the minimum requirements to be compliant with the Code. Despite the lack of incentives, the United Nations Conference on Trade and Development reported that, as a result of a survey, 64 percent of port operators perceived that the implementation of the Code had a positive effect since it offered a mechanism to standardize security in all parts of the port.

Supply Chain Security

Supply-chain security has not been globally mandated through the adoption of a code. As a result, it is governed by a number of programs, initiatives, and standards. Many of these approaches have different focal points in the implementation of security. However, they all have a built-in incentive for compliance that results in expedited entry for complaint entities. This approach encourages a more integrated adoption of security measures than those that merely satisfy regulatory compliance requirements and aid in making a compelling business case for full compliance. Some of the more widely known programs that influence supply chain security are discussed below.

Container Security Initiative (CSI)

The US CSI program, managed by Customs and Border Protection (CBP), involves bilateral agreements between the US and foreign governments for approximately 58 (as of May 2011) international ports from which goods are shipped to the US. CSI allows for the assignment of CBP personnel at the foreign ports of loading. These CBP personnel identify high-risk containers bound for the US that, if agreed to by host customs officials, are then screened (using x-ray and radiation-detection equipment) prior to loading. Once screened under the CSI program, the containers are typically not screened upon arrival in the US. The *quid pro quo* in the CSI arrangement is the expectation that CBP personnel will share intelligence with their host customs counterparts. It is also assumed that a close personal working relationship will evolve between the two customs services. A long-standing criticism of the program is that it is not unusual for host customs services to see the CSI program as largely "one-way," with them providing information and service to CBP but receiving little of value in return. Historically high turnover rates among CBP overseas-stationed agents have contributed to inhibiting the personal relationships between the two services' agents. CBP is currently developing procedures to stabilize assignment lengths at CSI ports. Host nations are under a formal obligation to inspect or screen cargo that CBP targets, and recent data shows that most requests are honored. In the event that the host country declines to screen, the cargo is targeted for screening upon arrival in the US. Further, as of 2010, there were no technical standards for screening equipment used by host-nation officials as part of CSI. While the program continues to expand, the lack of uniform participation of smaller ports that load cargo bound for the US has the unintended consequence of highlighting the vulnerability of smaller container ports for terrorist exploitation.

Authorized Economic Operators Guidelines (AEO)

In 2007, the European Commission issued guidelines for the issuance of an AEO Certificate based on demonstrable adherence to the security procedures outlined in the WCO SAFE Framework: financial solvency, transparency of ownership and management, and compliance with customs requirements. The incentives for certification include:

- Reduced information submission requirements
- Fewer operational controls
- Priority treatment if selected for controls

The AEO process is similar to the C-TPAT initiative, and negotiations are underway for mutual recognition between the two systems.

SAFE Framework

While CSI is a US-based inspection program, the World Customs Organization (WCO) developed the Framework of Standards to Secure and Facilitate Global Trade (SAFE Framework), which provides a mechanism for the development of international standards as well as the inspection of outbound cargo by customs agencies at the loading ports when requested by receiving nations. While the SAFE Framework has not become mandatory, 147 WCO members have committed to adherence to the standards. The SAFE Framework also provides a mechanism by which elements of the supply chain that demonstrate their commitment to adhere to the requirements as well as their integrity and transparency can receive preferable treatment, expedited customs clearance, and streamlined information submission requirements. Those entities can include "manufacturers, importers, exporters, brokers, carriers, consolidators, intermediaries, ports, airports, terminal operators, integrated operators, warehouses, and distributors."[1] While adherence to the requirements of the SAFE Framework is not mandatory, it is designed to provide tangible incentives to participants, unlike the ISPS Code, which is enforced through the fear of sanction or penalty.

Megaports Initiative

Since 2003, the US National Nuclear Security Administration (NNSA), a specialized agency within the Department of Energy, has provided nuclear- and radiation-detection equipment and training at large container ports in order to enhance the ability to detect attempts to smuggle radiological or nuclear material. In return, NNSA requests that the foreign government report any detections of material to NNSA. Under

---

FIGURE 5.3

Detection equipment at a megaport. *Photo credit: James R. Tourtellotte.*

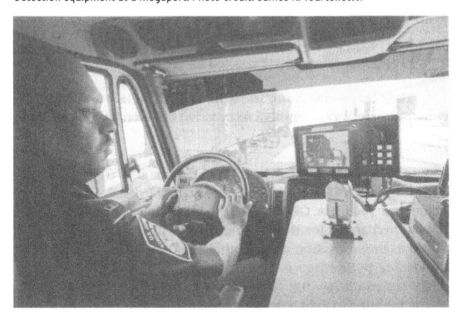

the Megaports Initiative, the NNSA provides equipment and training to host-nation officials, who then operate the equipment. In certain ports that are part of the Megaports Initiative, the NNSA coordinates its efforts with the US Customs and Border Protection.

International Port Security Program

In 2004, the US Coast Guard implemented a program to assess the compliance of foreign ports with the ISPS Code. This program involves deploying Coast Guard teams to audit the established ISPS program at selected ports. The results of the audits are used to assess the risk posed by ships coming from those ports. As part of the program, the Coast Guard assigns International Port Security Liaison Officers (IPSLO), who are responsible for particular geographic regions of the world and develop relationships with foreign agencies responsible for port security within their area of responsibility. These IPSLOs can also coordinate Coast Guard training for requesting nations. IPSLOs liaise with US Customs and Border Protection representatives when visiting CSI ports. While the Coast Guard has committed to working closely with CBP in the execution of the International Port Security Program, it is unclear whether regular information on risk is conveyed to CBP or other DHS entities from IPSLOs or foreign port audit teams. This is representative of a larger information-sharing problem surrounding cargo security across international borders (and within the US as well). Additionally, the audit teams typically visit a port for a short period of time and, due to the voluntary nature of the relationships between the IPSLOs and the host nation, are likely to be unable to accurately reflect the actual state of security in the ports rather than the ostensible state of security as represented by the host nation. Faced with the inability to accurately assess the level of integrity of the host-nation security officials through independent assessments, the value of the International Port Security Program audit function is limited.

Customs-Trade Partnership Against Terrorism (C-TPAT)

C-TPAT is a voluntary, incentive-based US program managed by CBP under which elements of the supply chain agree to follow certain guidelines. Participants are expected to maintain a specified level of security by adhering to those guidelines and may be audited by CBP. Eligible participants include US importers of record, US/Canada highway carriers, US/Mexico highway carriers, rail carriers, sea carriers, air carriers, US marine port-authority/terminal operators, US airfreight consolidators, ocean transportation intermediaries and nonvessel operating common carriers (NVOCC), Mexican and Canadian manufacturers, certain invited foreign manufacturers, and licensed US customs brokers. Other supply-chain elements can participate if invited by CBP. As an incentive, participants receive priority processing at the border and CBP training opportunities. The fundamental C-TPAT challenge is its voluntary nature. For political reasons (including budgetary constraints) mandatory security standards have not been promulgated. The C-TPAT guidelines are fundamentally sound, but since the nation's principal cargo border-security support program consists of voluntary guidelines, with punishment being removal from the C-TPAT participant list and delay at the border, the accountability for cargo security rests entirely with government and its screening capability instead of being shared between government and the private sector. A potential strength of the C-TPAT program is the expedited movement of cargo belonging to qualified participants. However, research is inconclusive as to whether companies are realizing a meaningful financial advantage from participating when compared to the costs of being approved.

International Organization for Standardization

International Standard 28000:2007 Specification for Security Management Systems for the Supply Chain (ISO 28000) is particularly beneficial to maritime security because it provides general guidance on the implementation of supply-chain security measures while also including a port-security standard, ISO 20858:2007 Ships and Marine Technology Maritime Port Facility Security Assessments and Security Plan Development. However, ISO 20858 does not include a standard for ship security, which results in a gap. Another subsidiary standard to ISO 28000 is ISO 28001 2007 Security Management Systems for the Supply Chain Best Practices for Implementing Supply Chain Security, Assessments, and Plans Requirements and Guidance. ISO 28001 amplifies the general approach of ISO 28000 by providing more detailed guidance in the implementation of supply-chain security measures. ISO 28001 specifically endorses the EU's Authorized Economic Operator concept.

Lack of Recovery Planning for Key Maritime Supply Chain Components

While there are national recovery plans and requirements for regional recovery plans for maritime infrastructure, there is no recovery plan that deals exclusively with the reconstitution of border and cargo security capabilities. Recovery considerations should include the ability to reconstitute border and supply-chain security capacity, the ability to implement alternative border-control facilities and/or procedures, and the need to prioritize border-crossing traffic and types of cargo. In the event of the sealing of US borders or an unexpected lack of access to certain border crossings, seaports, and/or airports due to either man-made or natural disaster, plans need to be in place to either re-open the affected port of entry or to develop alternatives. This is particularly important due to the "just-in-time" delivery nature of modern logistics. The impact of a slowed or stopped logistics chain can be economically corrosive, as evidenced by a 2005 study in which both US and Canadian shippers were polled regarding the impact of increased US border-security controls. In that study, border-crossing delays caused over 40 percent of American importers and almost 35 percent of Canadian importers to consider measures to reduce their respective dependence on imports from the other side of the border.[2] The actual shutdown of ports of entry can have catastrophic economic effects. A recent example is the West Coast longshoremen's strike of 2002, in which estimated economic losses were as high as $1 billion a day. Additionally, the criticality of ports of entry can vary due to season or other regional economic considerations. For example, ports in New England that receive home-heating fuel are far more critical to the regional population in January than they are in July.

A Disjointed International Regulatory Environment

As a result, most of the wide variety of national and international codes, regulations, standards, and programs have both strengths and weaknesses. As will be shown in future chapters, an approach in which port, ship, and supply-chain security are consolidated into a single, internationally accepted standard will address most of the weaknesses identified in the current regulatory environment.

Overreliance on Technology

In an effort to develop efficient and foolproof security measures, there has been, over the last decade, an increased focus on the use of technology in security, especially in the areas of access control and detection. While this is a natural development that reflects a desire to take full advantage of rapidly changing and developing technologies, it can provide a false sense of assurance that the deployment of technology can serve as a substitute for effective planning, good procedures, and the decision-making that can only be made by professionals with knowledge and experience in maritime security.

Part of the rationale for an increasing focus on technological security solutions is the perception that they are more cost-effective and reliable than traditional security measures involving the use of security officers to control access and secure facilities. Further, the extensive use of technology in established, fixed, and predictable ways may in fact create additional vulnerabilities by eliminating or reducing the random security activities that have been known to deter criminals or terrorists. In other cases, however, technology that is deployed based on well-crafted system requirements to accomplish specific objectives can effectively enhance security as part of a multidisciplinary maritime-security program. The following are examples of technological solutions that were either proposed or used in ways that did not serve to provide or enhance security.

Maritime Domain Awareness (MDA)

Maritime Domain Awareness is a term that is defined by the United States as "the effective understanding of anything associated with the global maritime domain that could impact the security, safety, economy, or environment of the United States."[3] Maritime domain is defined as "all areas and things of, on, under, relating to, adjacent to, or bordering on a sea, ocean, or other navigable waterway, including all maritime-related activities, infrastructure, people, cargo, and vessels and other conveyances."[4] Therefore, MDA appears to be an effort to gain complete and transparent understanding of all activities in all coastal areas and the high seas throughout the world. The intent is that achieving MDA will allow governments to take early or preemptive action to mitigate threats or to be fully prepared to effectively respond to them. While this is both a noble and ambitious concept, implementation thus far has been primarily focused on the acquisition of sensor equipment and the development of fusion capabilities, especially the development of a common operating picture (COP). The COP approach is sensor-intensive and can be valuable in monitoring real-time maritime traffic. With the exception of the automated information system (AIS), which is easily manipulated or turned off, most of the sensors appear to be reliable. AIS was designed for maritime-safety purposes such as facilitating search and rescue. It is required on most ocean-going commercial vessels on international voyages but is not designed for security purposes. Therefore, it can be manipulated by on-board personnel, and the identifying information, intended ports of call, and other key information can be changed or deleted. Further, the system can be completely turned off, rendering AIS fundamentally unreliable.

The biggest challenge facing MDA is how to effectively utilize a system designed to collect massive amounts of tactical information into a tool for real risk-based decision-making.

FIGURE 5.4

A shipping container being scanned.

The Fallacy of 100 Percent Container Screening

A good example of the proposed use of technology to enhance security that may, in fact, decrease the efficacy of random security measures is found in the proposal to require all containers bound for the United States to be screened prior to being loaded onto ships at their port of departure. This proposal is neither effective nor necessary and may cause trade with the United States to become more expensive while not significantly reducing the risks that the regulation is intended to address.

In 2006, the US Congress passed legislation (Public Law 109-347, The Security and Accountability for Every Port Act of 2006, or SAFE Act) requiring that 100 percent of all containers be screened prior to being loaded on ships bound for the US. The containers were to be screened for radiation, both shielded and unshielded. The implementation of this requirement caused significant concerns within the private sector about the cost and speed of importing goods into the United States, especially as the concept of just-in-time delivery continued to take root. The impetus behind the perceived need to impose 100 percent screening is the assessment that containers are a potential way to smuggle contraband, especially radiological or nuclear weapons of mass destruction (WMD). As will be shown later in this chapter, the lack of a focused consideration of the threat and the resultant risk of the scenario by political figures (who coincidentally may have been interested in being perceived as advocating strong, easily understood security measures) likely contributed to the decision to pursue 100 percent container screening.

The real risk of a WMD being injected into the commercial supply chain in a container is relatively low given the behavior of the terrorist groups that would most likely obtain such a weapon. This will be further explored in following sections of this chapter. As a result, the proposed regulation would increase the costs of both the US government, which would have to assist in the funding of technology in some foreign ports and hire additional personnel to coordinate the program, and industry. The private sector would be saddled with possible additional delays in shipping, which would increase operating costs while realizing no incentives for complying if other global markets without the requirements could be identified, thus placing US businesses and consumers at a competitive disadvantage.

Fortunately, in June 2011, the Secretary of Homeland Security announced that 100 percent container screening was not a realistic goal, there would not be an attempt to institute the requirement, and other intelligence, risk, and incentive-based programs would be used to target suspect containers and to facilitate trusted shipments.

The "Magic" of Closed Circuit TV (CCTV)

Since the attacks of September 11, 2001, the deployment of CCTV systems in ports and on ships in the US has increased exponentially. This has been encouraged by the increased use of CCTV throughout the security industry, as well as the highly publicized systems installed in London and New York City. The propensity to install CCTV systems has been further exacerbated by the US Port Security Grant Program, which for many years has been oriented towards technological procurement initiatives rather than training and staff hiring.

However, the most egregious problem with CCTV is the common lack of understanding regarding its utility as part of a security system. While CCTV installation may have some deterrent value to potential criminals and (less likely) terrorists, it is mostly effective for investigative use and minimally useful for actually responding to or preventing security incidents.

London is often cited as the most heavily covered by CCTV city in the world, which as been the case since the early 2000s. In July 2005, London was attacked by Al Qaeda operatives who targeted the Underground. One of the operatives also attacked a bus as a secondary target. The attackers were captured by CCTV systems on numerous occasions on the day of the attack as well as on reconnaissance trips prior to the actual attacks. The extensive CCTV system had no effect in either preventing or detecting the attacks, and its utility was only in the subsequent investigation. Further, the camera system did not serve as a deterrent to the attackers. This example amply demonstrates the limited effectiveness of a CCTV system. A response capability is only possible when there is a response force that is on constant alert within the immediate area of the CCTV system cameras. One area of value that CCTV systems bring to the ability to detect security events in real time is the ability to "diagnose" the nature of a threat and to take precautionary measures to reduce the impact of the security event.

Of particular concern with the procurement of technology is the belief that the use of technology will be cost-effective and the common failure to program funds for maintenance, training, repairs, and upgrades beyond the initial procurement and installation costs. This was very evident in many rounds of the US Port Security Grant Program, which did not allow grant monies to be spent on system sustainment.

Failure to "Fire for Effect"

In both of the previous examples, the key problem was a failure to select appropriate technological solutions based on system performance requirements that were not clearly articulated. In the case of the US attempt to impose 100 percent container screening, the system mission should have been to prevent WMD from entering the US. However, the system requirement was either not fully articulated, based on a faulty risk assessment, or was focused on detecting the importation of WMD vs. preventing WMD from entering the country.

In the case of the London CCTV system (and almost every CCTV system deployed), an effective development of a system would have led to the determination that CCTV systems alone, will not prevent an attack and do not contribute significantly to the immediate response to an attack while it is occurring.

The Staten Island Barge Explosion

Despite the previous examples of CCTV being improperly utilized, there are occasions where CCTV has been useful in supporting the rapid assessment or diagnosis of events in order to determine preliminary causes and the level of initial response. An excellent maritime example can be found in the use of CCTV in assessing the initial cause of a fuel barge explosion on February 21, 2003, while moored at a terminal in Staten Island, NY. The explosion was felt throughout Staten Island, and the smoke was seen throughout New York City and northern and central New Jersey, for whose residents the memories of the attacks of September 11, 2001, were relatively fresh. A CCTV camera from the US Coast Guard's Vessel Traffic Service in New York Harbor captured the explosion, and rapid review of the footage provided sufficient information to determine that the explosion was not likely due to a terrorist event. The determination that the explosion

FIGURE 5.5

A video still of the barge explosion on Staten Island, NY, in February 2003. *Photo credit: US National Oceanographic and Atmospheric Administration.*

was not related to terrorism resulted in the security of the port not being elevated to a level that would impede or shut down commercial operations.

Minimizing the Importance of Understanding Threat

As one of the key components of risk, along with vulnerability and criticality, threat may be the least understood, which contributes to a tendency on the part of some security professionals to minimize the importance of understanding the threat to a particular facility, organization, or other entity. Particularly in the United States, there has been a movement to focus on "all-hazard" risk assessments, which has served to further dilute the importance of developing a context in which to consider the issues of vulnerability and consequence. Without a well-defined threat assessment, security risk assessments cannot adequately determine the vulnerability of the assets, facilities, or organizations to the harm or the likely consequences of any harmful acts.

Hurricane Katrina–the Wrong Lesson Learned

For several years after the attacks of September 11, 2001, the US approach to security risk did place an emphasis on security threats to potential targets. While this emphasis did not necessarily provide for a sophisticated threat-assessment model or methodology, it did exist. However, in 2005, a key agency within the US Department of Homeland Security, the Federal Emergency Management Agency, was criticized for the largely incompetent and ineffective coordination of the response to and recovery from Hurricane Katrina on the US Gulf Coast. The massive criticism of FEMA and resulting takeover of the response by the US Coast Guard and other armed forces due to the inability of civil authorities at all levels of government to effectively handle the crisis created a reactionary impetus within the Department of Homeland Security to be prepared for both security events as well as large-scale natural disasters and accidents. This shift in focus led to an approach that emphasizes response preparedness and rapid recovery and places less of a priority on the identification and prevention of harmful acts before they occur. This shift was cemented by the introduction of an "all hazards" approach to risk in which the reasons an event happen are less important than their impact. The nature of this approach deemphasizes the external actors in a potential event and emphasizes the assessment of vulnerabilities and consequences. The problem with this approach is twofold. First, the "all hazards" approach, while acknowledging the need to provide context to the potentially infinite vulnerabilities and consequences that could be identified through the development of scenarios, does not emphasize the development of threat scenarios by potential threat actors that are tailored to the facility, entity, or organization, being assessed. The generic nature of "all hazard" assessments may not address the specific threats that the assessed target may face. Second, due to the lack of a rigorous assessment methodology that would otherwise be found in a properly developed threat assessment, the identification of hazards and their scenarios tends to be minimized and the greater emphasis is placed on vulnerability and consequence possibilities, which can be almost infinite. Finally, the "all hazards" approach deemphasizes the detection and prevention elements of security risk and emphasizes the response and recovery aspects. This approach has basically turned security risk management from a multidisciplinary

FIGURE 5.6

A flooded New Orleans.

way of ensuring the security and resilience of the protected asset to an emergency-management approach that only addresses one portion of the full security spectrum of activities.

Assessing Threat is Hard

Assessing threat is not an exact science or art. As noted in earlier chapters, the capabilities and intentions of human adversaries cannot be reasonably calculated with certainty. Operations research professionals and consultancies have made attempts to calculate threat and risk, but these methodologies have no grounding in the realities of dealing with an adaptive, intelligent enemy. The work done by operations researchers and others seem to focus on probability and statistical analysis and to exclude the value systems, tactics, and adaptive capabilities of terrorist and criminal groups. While other hazards such as natural disasters and accidents can be assessed using past data and establishing likelihood and probability, this is not true for humans who have nefarious intentions. In the United States, as the emphasis has continued to focus on "all hazard" approach, the ability to predict the likelihood of natural disasters and accidents through statistical analysis provides some degree of comfort to policy-makers and others responsible for preparing to respond to crises. Therefore, the fact that accurate threat assessment is made more difficult by the "cat and mouse" approach and ambiguity of trying to predict and anticipate an intelligent adversary, coupled with the inability to effectively calculate the likelihood of an attack or criminal act, makes any attempts to assess threat suspect in the eyes of bureaucrats more used to planning response and recovery operations against less ambiguous challenges.

Why Understanding Threat Matters

Security practitioners cannot design effective security and resilience measures if they don't understand their adversaries. A rigorous threat assessment provides the key context in which to plan the full spectrum of security operations, including detection and prevention measures. It also provides efficiencies in the security risk-management efforts by providing more realistic possible security events to be planned against rather than a broad, amorphous set of generic scenarios. Therefore, if a well-crafted, rigorously analyzed threat assessment is not included in a security risk assessment, that assessment lacks fundamental credibility, as it will be focused on vulnerability and consequence in which the possible threat or hazard scenarios are unlimited. Security measures cannot be put in place to protect an entity in which everything is vulnerable. Further, an understanding of specific threats, tailored to the specific asset being assessed, will provide the key components in the development of design basis threats, which in turn will drive the development of effective and targeted risk treatments. A perfect example of a security problem that did not reasonably consider threat is the case study involving 100 percent container screening that was mandated in the United States.

Bomb in a Box?

As noted previously, in passing the SAFE Act of 2006, the US Congress legislated a requirement for 100 percent of all containers destined for American ports to be screened at the ports of loading outside the US. The specific requirement is for the ability to screen all containers for radiation, thus establishing that the concern was for the potential smuggling of radiological or nuclear material. This requirement is a classic example of a risk-averse vs. risk-based approach to risk management. Specifically, the premise that there is a likely threat of the smuggling of radiological or nuclear material via a commercial container is not based on a well-developed or analytically rigorous threat assessment but falls into the expansive category of "It could happen." If the proposition that a terrorist group that obtains a nuclear weapon would smuggle it into the US via a container, is viewed through the prism of an analyst who understands the threat posed by terrorist groups, including their tactics and history, the likelihood of the smuggling of a nuclear weapon in normal, commercial container shipping is significantly less likely. Part of this may be explained by the phenomenon known as the "social amplification of risk," where popular perceptions of risk do not match the actual risks of a particular event happening.

Deconstructing the Threat

While it is clearly an objective of Al Qaeda and other terrorist groups that include mass-casualty attacks as methods to smuggle weapons of mass destruction (WMD) into the United States, the mode of transport and delivery is a subject that is ripe for debate and discussion. Al Qaeda has publically embraced the concept of attacking Western targets with WMD since 1998 and has shown significant interest in acquiring the capability to execute a WMD attack. The options for Al Qaeda in developing a WMD are limited.

Biological and Chemical Agents

While biological and chemical agents could be used in an attack, many of the ingredients necessary to develop them are available in the United States, making the requirement for smuggling from abroad largely unnecessary.

Radiological Material

Radiological material, necessary for the construction of a radiological dispersal device or "dirty bomb" can also be found in the United States, thus lessening the likelihood of needing it to be smuggled. Material can be found in common applications in American commercial, medical, and research facilities. The explosives necessary to disperse the radiological material are even more easily obtained.

The materials needed to develop or construct biological, chemical, or radiological devices are largely available in the United States, thus likely precluding the need to smuggle them into the country. It is possible, however, that a more potent or ready-made foreign source could be smuggled into the country, but it is unlikely and certainly not a requirement.

The Nuclear Grail

Therefore, the most likely WMD that would originate outside the United States would be a nuclear weapon. Al Qaeda has clearly shown an interest in obtaining a nuclear weapon either through building it or obtaining it from existing stockpiles, most likely in the former Soviet Union or Pakistan. If Al Qaeda or a similar group were to obtain a nuclear weapon, there would clearly be interest in detonating it in the United States or other western country. Since the weapon would almost certainly be procured outside the United States or other western target country, it would have to be smuggled into the location where it would be detonated. The issue becomes the level of control that Al Qaeda is willing to surrender to commercial shipping interests by placing their most valuable weapon into a container that then leaves their control and is loaded on a ship bound for the United

States. Shipping containers can become damaged, lost, or redirected. Further, the ships that carry them can suffer engineering casualties or be diverted due to weather or changes in schedules. Containers can even be destroyed or washed overboard in storms. All of these possibilities could result in the loss or seizure of the WMD due to the vagaries of international maritime commerce. Therefore, the fundamental issue for consideration is whether Al Qaeda would relinquish control of its "Holy Grail" in order to deploy it or whether it would maintain control of the weapon and find another way to bring it to the targeted location. Generally, the assessment by intelligence analysts and other terrorism experts is that Al Qaeda is unlikely to simply insert a WMD, specifically a nuclear weapon, in to the normal container supply chain.

The Risk Conundrum

Based on the unlikely scenario of a WMD of any sort being transported via container, the requirement for 100 percent screening of containers seems unreasonable and designed to mitigate any risk. Clearly, due to the fact that the possibility of a smuggled WMD cannot be eliminated, some sort of attention needs to be given to container traffic in addition to the possibility of detecting other contraband. The issue is whether an investment in a program geared only to

---

FIGURE 5.7

Container ship with storm damage and missing containers. *Photo credit: US Coast Guard.*

radiation detection and requiring 100 percent screening is a properly considered risk-based approach or whether there are more cost-effective processes that may be more effective against a broader suite of threats. What is clear is that without a fundamental understanding of the threat and the adversaries' thought processes and approach, the debate is fundamentally flawed.

The Consequences of not Understanding the Threat

In some cases, not understanding the threat can lead to reactions or responses that are completely inappropriate and don't focus on the actual threat actor. An example of responding to the wrong threat is the US invasion of Iraq in 2003.

Hitting the Bystander

Regardless of the potential political and ideological reasons that were behind the invasion of Iraq in 2003, any student of Al Qaeda at the time understood that not only was the Baathist regime of Saddam Hussein not an ally of Al Qaeda, but the two were in fact enemies and adherents of incompatible ideologies. Shortly after the attacks of September 11, 2001, there was much public discussion regarding Iraqi support for Al Qaeda as a rationale for invading Iraq. This was reinforced by the Iraqi regime's refusal to acknowledge the state of its WMD research program or stockpiles. While neither the Baathist regime in Iraq or Al Qaeda are deserving of sympathy, the fact remains that the rationale for invading Iraq was based on a fundamentally flawed assessment of the threat to the United States by Iraq.

Al Qaeda's View of Saddam's Iraq and Vice Versa

Al Qaeda's view of Iraq was that of an "apostate regime" that was focused on secular governance and was not following Sharia law. The only point of support that Osama bin Laden showed to the Iraqi regime of Saddam Hussein was in December 2004, after the invasion, when he stated that Saddam was the lesser of two evils. Saddam's view of Al Qaeda was one in which he considered the organization to be led by a "zealot" and that the Baathist ideology included a belief in the separation of religion and government. Saddam also noted that, while he opposed some US policies, he did not consider the US to be an enemy, while Al Qaeda clearly did. Therefore, there was no incentive for Iraq to support Al Qaeda's plans to attack the United States.

Finally, Saddam did not cooperate with the efforts to determine whether Iraq had WMD primarily because he was worried about his real enemy, Iran. Saddam, in interviews prior to his execution, stated that he was worried about appearing vulnerable to the Iranian regime, Iraq's enemy from a bloody war in the 1980s and the source of potential unrest in Iraq's Shia communities.

The Threat That Wasn't

While the Iraqi government under Saddam Hussein was brutal and oppressive, it didn't ally itself with Al Qaeda or support the attacks of September 11, 2001. However, a significant political and military effort to justify and execute the invasion of Iraq was based on a flawed understanding of the threat posed by Al Qaeda.

The Fallout

The poor analysis and/or political objectives that resulted in a false assessment that Iraq was a threat to the United States had consequences that go beyond the lives destroyed, money spent, and upheaval generated in the region. It can be argued that the invasion of Iraq actually improved Iran's presence in the region and provided it with a presence in Iraq and its Shia population that previously did not exist.

The Lack of a True Risk-Based Approach

As noted in previous chapters, there are two accepted definitions of risk in maritime security. The US version used by the Department of Homeland Security defines risk essentially as the potential for harm, while the definition most accepted outside the United States defines risk as the chance that something may happen to affect objectives. Since the US definition is expressed in such a way that risk is inherently negative, the US approach to risk is to reduce or mitigate rather than treat. This rigid approach leads to fewer options to address risk and inhibits a truly risk-based approach. An example can be found in the FEMA's fiscal year 2012 *Port Security Grant Program (PSGP) Investment Justification Template.*[5] In the template, which provides the key justification requirements and criteria for the grant program, applicants are asked to address how their proposed project would "reduce risk in a cost-effective manner." The template does not ask applicants to identify how the potential project assessed risk or how the applicants may have prioritized risk. This approach reinforces the approach that risk should only be mitigated or reduced and ignores any efforts to prioritize risk by developing risk registers or determining the risk tolerances of stakeholders and infrastructure or system operators.

Finally, in the United States, risk assessments pertaining to maritime security and arguably homeland security writ large are artificially limited by a number of federal bureaucratic entities that have jurisdiction over discrete geographic areas that follow political boundaries but not economically or culturally significant or interrelated locations. These bureaucratic boundaries lead federal and state agencies to

FIGURE 5.8

The New York-Newark-Bridgeport, New York-New Jersey-Connecticut-Pennsylvania combined statistical area.

FIGURE 5.9

The Bridgeport-Port Jefferson Ferry route.

___

potentially overlook the possible broader interdependencies within larger metropolitan areas or the possible cascading consequences of a crisis in one jurisdiction spilling over in to a neighboring area.

For example, the New York City metropolitan area, identified by the US Office of Management and Budget as the New York-Newark-Bridgeport, New York-New Jersey-Connecticut-Pennsylvania Combined Statistical Area, includes 30 counties in four states.

Within this region, interconnected by economies, neighboring jurisdictions, transportation systems, and a large commuting population, are numerous federal agencies whose jurisdictions only cover part of the region, thus preventing a broader and more effective risk-based approach to maritime security. For example, parts of the waterways and ports within the Combined Statistical Area fall under the purview of three different US Coast Guard Captains of the Ports, two FBI Field Offices, and two FEMA Regions. This bifurcation of an integrated geopolitical area inhibits a fuller understanding of the potential impacts of risks throughout the metropolitan area.

CASE STUDY

The Bridgeport-Port Jefferson Ferry operates large car and passenger ferries between Bridgeport, CT, on the US mainland and
Port Jefferson, NY, on Long Island, with yearly passenger traffic
averaging over 1 million and 450,000 vehicles. What is of significance is that on September 11, 2001, and into September 12,
the ferry line was one of two connecting the 7.5 million residents
of Long Island, NY (comprising two suburban counties and the
New York City boroughs of Brooklyn and Queens) with the US
mainland. This was because the bridges and tunnels linking Long
Island with Manhattan, Staten Island, and the Bronx were all shut
down for almost 24 hours for security reasons and to facilitate
access to the site of the attacks by first responders. Therefore,
the ferry system clearly showed its value to the metropolitan
area and Long Island in particular. Further, the Bridgeport-Port
Jefferson Ferry is also one of the very few privately owned and
operated, for-profit ferry lines in the United States, and therefore
all costs associated with enhancing security, with the exception
of grant awards, are overhead. Other ferry systems that are gov-
ernment- or quasi-government-owned and -operated enjoy government subsidies as well as the opportunity to compete for the
same grants as the Bridgeport-Port Jefferson Ferry system. This
lack of recognition as a vital component of regional resilience is
likely a result of the ferry system crossing several jurisdictions. It
operates in two states, across two FEMA regions, but serves one

Insufficient Focus on System Integrity

A major issue affecting current efforts to enhance international maritime security is that of a lack of transparency in the implementation of security measures and the varying levels of corruption around the world. The lack of transparency and potential for corrupt practices that may undermine security efforts mean that many international security measures are possibly far less effective than designed. Obviously, due to the global nature of shipping, if one element is compromised, the whole system is made more vulnerable.

Transparency

For the purposes of this book, transparency is defined as openness or accountability. A lack of transparency means that key decisions are not made in the open and that the accountable parties for maritime security are not clearly defined. Therefore, the less open the process is regarding the implementation of security measures or the less accountable responsible parties are for their implementation, the less confidence there is in the ability of the system to work properly. In maritime security, one of the most important yet least transparent influences on supply-chain security measures is their effect in reducing insurance costs. The insurance industry is extremely reticent about releasing information regarding how the decisions about insurance rates to ships, port, facilities and other elements of the maritime domain are determined and how security measures may affect those rates.

Corruption

Corruption is the willful circumvention of maritime-security measures by those responsible for either enforcing or executing them, usually for gain or profit. Like transparency, this can affect both the public-sector agencies and the private-sector operators. Corrupt public-sector representatives can allow illegal people, goods, weapons, or other contraband into the secure facilities or port areas under their jurisdiction. Corrupt private-sector operators can facilitate the circumvention of maritime, port, and supply-chain standards.

Implications for the Maritime Domain

A country or port whose officials and associated industries are assessed as being corrupt should not be considered fully compliant with international maritime security measures. Further, a low degree of transparency should lower the degree of confidence that a country or entity is carrying out security measures as advertised. This is exacerbated by the fact that ISPS is a nationally self-regulated program in which national administrations implement the Code and then report their level of compliance. There is no internationally required assessment process whereby third-party, neutral assessors determine the level of actual compliance with the ISPS Code. The assessment function is different from the audit function, which is found in many international standards but not in the ISPS Code. Audits are conducted to determine the compliance of the audited party with the implementation of a standard, the "mechanics" of the standard. The US Coast Guard has an International Port Security Program that audits nations' compliance with the ISPS Code, but the teams usually spends a relatively short period of time in the audited ports and is not charged with identifying corrupt practices among audited parties. Further, supply-chain security standards such as the SAFE Framework and ISO 28000 are voluntary, and, while there are audit features, they are also not focused on the integrity of the system. Therefore, there is a strong likelihood that, while countries will report that they are compliant with the ISPS Code or committed to implementing the WCO SAFE Framework, their record of corruption or their lack of transparency suggests that their compliance level is in reality questionable. This lack of confidence, if known, would likely create a level of uncertainty or doubt on the part of trading partners and thereby negate confidence in the actual effectiveness of international standards and codes. Figure 5.10 shows the most corrupt nations (with maritime industry) as assessed by Transparency International, an international watchdog of corruption, and the self-identified compliance with ISPS and whether they are a signatory to the WCO SAFE Framework.

The Impact of Corruption

The best-documented security processes or physical security measures are irrelevant if circumventable by public officials or the industry. The following figure demonstrates the uncertainty of self-reporting if countries assessed as being the most corrupt in the world report that they have implemented ISPS. If there is no external, neutral verification of proper implementation, then the entire ISPS system's integrity is questionable.

Country	Corruption Rank[1]	Self Report ISPS Compliance[2]	WCO Framework[3]
Angola	168 of 182	Yes	Yes
Libya	168 of 182	Yes	Yes
Equatorial Guinea	172 of 182	Yes	No
Venezuela	171 of 182	Yes	No
Haiti	175 of 182	Yes	Yes
Iraq	175 of 182	Yes	No
Sudan	177 of 182	Yes	Yes
Myanmar	180 of 182	Yes	Yes
North Korea	182 of 182	Yes	No
Somalia	182 of 182	No	No

1   http://cpi.transparency.org/cpi2011/results/. Accessed 14 April 2012
2   http://gisis.imo.org/Public/. Accessed 14 April 2012
3   SUPPLY CHAIN SECURITY: CBP Works with International Entities to Promote Global Customs
     Security Standards and Initiatives, but Challenges Remain,
     http://www.gao.gov/new.items/d08538.pdf. Accessed 14 April 2012.

FIGURE 5.10

The most corrupt countries with self-reporting of compliance with ISPS.

Lack of Incentives for the Private Sector

While some security programs, such as the Container Security Initiative, CT-PAT, and WCO SAFE Framework, are designed to provide incentives for the private sector to comply, other programs or requirements, such as the ISPS Code, do not have incentives built into their implementation. This is largely due to the lack of transparency in the maritime insurance industry and an inability to determine how beneficial security measures could be in lowering insurance rates. While the ISPS Code is mandatory in most countries, the lack of an incentive program does not encourage any behavior beyond basic compliance. Further, this lack of transparency among insurance providers makes it difficult for security managers and government officials to make a business case for enhanced security measures, with the exception of those supply-chain programs that offer preferred and expedited treatment for full compliance.

REFERENCES

1. World Customs Organization. WCO SAFE Framework of Standards, 2007, p. 6.
2. http://www.importers.ca/industry_news/2005/05_02_24_impacts_us_govt_antiterrorism.pdf. Accessed on October 25, 2007.
3. National Plan to Achieve Maritime Domain Awareness for the National Strategy for Maritime Security. October 2005, p. 1.
4. Ibid.
5. http://www.fema.gov/pdf/government/grant/2012/fy12_psgp_invest.pdf. Accessed on 14 April 2012.

PART FOUR

Principles for Effective Maritime and Port Security

CHAPTER 6

Security as an Enabler

INTRODUCTION

Effective maritime security cannot be considered in isolation. Security affects and is impacted by numerous external factors, including socioeconomic drivers, political priorities, transportation-system linkages, business trends, and international events. Therefore, a truly risk-based approach to security is imperative. Further, because of the complex interactions of ports, ships, and the maritime domain with other economic interests, logistics, and transportation modes, security must be considered to be an element of system resilience and risk management. Security of the maritime domain and its associated systems cannot be, by itself, a primary objective but should be considered within the context of ensuring safe and efficient maritime operations and commerce. Properly and expertly designed security strategies and measures serve as enablers to allow for the continued cost-effective and reliable operation of industries, government services, and economies. While security is often considered a cost center in the commercial environment, an integrated approach can ensure that security and resilience serve to minimize the costs of disruption while maximizing the reliability and competitiveness of business operations. This approach may minimize the burdens of security in a competitive business and economic environment while enhancing system reliability. An approach integrated with resilience and sensitive to the needs of businesses and government can best be achieved through an outcomes-based approach in which tailored and imaginative processes can be used to realize the desired security and resilience objectives. Prescriptive processes should only be required where absolutely necessary.

Why is it Important for Security to be an Enabler?

Taking an integrated approach where security is seen as an enabler rather than an objective will serve several fundamental purposes that will make the mission of an effective maritime-security practitioner to protect the assets or domain that he or she is responsible for more achievable:

- Security will eventually be perceived as less of a cost center for businesses and will be seen to add value.
- An integrated approach will help facilitate a culture of security throughout the organization.
- The security function will likely be perceived as less heavy-handed in cases where that is a concern.
- Security will be understood to be a vital element of corporate or agency success.

Security as a Value-Add

Reducing security's function as a cost center to an enterprise can be accomplished by the integration of security into the operational functions of an organization. Examples of this can include demonstrating the value of the security department in ensuring that suppliers are able to protect their shipments through supply-chain security initiatives and supplier site visits and developing measures of effectiveness that show the positive impact on efficiency and the reduction of organizational delays (truck and container movements, ship arrivals, reduction in cargo theft, etc.) that can be attributed to security measures.

A Culture of Security

By coordinating security with other facets of an organization's operations, especially environment, health, and safety (EHS), there is an economy of scale in communicating risks and policies throughout the organization. By also including security communication and awareness training in other venues that are not necessarily security-specific, there are opportunities to show security's role as a key and integrated component of an enterprise's functions.

Changing Security's Image

Security practitioners and the security discipline in general sometimes suffer under stereotypes that include perceptions of being heavy-handed, preventing the free flow of information, goods, and people, and generally impeding business or operations. By integrating the security process with other disciplines where possible, the ability to change the misconceptions about security can be addressed. Additionally, a regular communications policy in which security measures are explained and feedback is requested will help reduce the mystery of the security function.

Security a Key Organizational Component

By pursuing an integrated approach to security, it can be understood as a key and vital component of any organization or entity, and the

---

FIGURE 6.1

A common stereotype of security officers.

security function can be presented as a valuable component of protecting the viability of an organization through its protection of life, property, and goods as well as its protection of supply-chain integrity and intellectual property. This integrated approach to security is rooted in the concept of system, port, and vessel resilience.

Resilience

Resilience with regard to critical infrastructure is defined by the American Society of Civil Engineers as "the ability to withstand and recover from extreme conditions."[1] By addressing the resilience of maritime transportation systems as an element of critical infrastructure, the vulnerabilities and possible consequences of natural disasters, accidents, or terrorist or criminal activity can be effectively managed and mitigated. This approach further enhances the effectiveness of a risk-based approach to maritime security, which is predicated on the ability to treat risk by managing changes in consequence and vulnerability. Further, resilience is focused on maintaining the ability of the protected system to continue operations in all but the most extreme conditions.

FIGURE 6.2

Prestaged emergency supplies. *Photo credit: FEMA.*

Examples of resiliency measures in transportation systems include redundant transportation routes, hubs (nontraditional ports and airports that could be used in emergencies to maintain the flow of commerce and critical supplies), diffused decision-making authority, robust planning, identification of alternate operating sites for management, focused physical security on single points of failure, regulatory flexibility in emergencies, and prestaging of critical supplies.

Why Resilience?

Building resiliency into transportation systems and their associated infrastructure is advantageous for numerous reasons. However, the most important reason for designing a security program that is consistent with a resilience-focused approach is the risk in ignoring or dismissing resilience.

Risks of Ignoring Resiliency

Several studies have investigated the potential risks posed by organizations, whether public or private, that do *not* have a mature resilience program in place. Some of the key issues that document a high level of exposure include:

- Likelihood of a crisis—research conducted of a sampling of large multinational firms shows that within any given five-year period, there is an 80 percent likelihood of a major crisis occurring that will significantly impact operations.
- Stock declines—stock prices are sensitive to crises and trade disruptions. Research among the multinationals has revealed a minimum 20 to 30 percent decline in corporate stock price in the month following a significant crisis. The research also shows that the performance of corporate executives in managing a crisis is the key determinant of whether a corporation rebounds from crisis or continues in decline. Roughly a fourth of the surveyed companies took more than 12 months for their share prices to recover.[2]
- Inherent risks in globalization—as firms expand globally, they are exposed to additional complex risks. Extended

supply chains, technology interdependencies, IT vulnerabilities, uncertain geopolitics, and even natural disasters all combine to increase the amount and complexity of business risk.[3]

- Corporate failure—of all businesses that close down following a disaster, more than 30 percent never reopen again.[4]

- Negative financials—analyzing 800 incidents of supply-chain disruption, it was assessed that companies that suffer disruptions experience between 33 to 40 percent lower stock returns relative to their benchmarks, an average of 13.5 percent increase in share price volatility, a 107 percent reduction in operating income, lower sales growth, and an average of 11 percent increase in costs. This does not include the amount of time required to recover the losses if recovery is possible.

FIGURE 6.3

A damaged business in New Orleans after Hurricane Katrina. *Photo credit: US Navy.*

Additional Risks

Negative performance metrics of disruptions compared to the year prior to the disruption often continue for up to two years after the disruption and can include the following:[5]

- Up to a 107 percent drop in operating income
- A 114 percent drop in return on sales
- A 93 percent drop in return on assets
- A 14 percent growth in inventories

The Benefits of a Resilience Approach

The benefits of taking a resiliency-focused approach to transportation security include:[6]

- Assuring system survival under adverse conditions by planning to sustain core operations
- Decreasing operational expenses and liabilities through lower insurance costs (where possible) and litigation costs, decreased theft, reduced employee turnover, and increased competition among suppliers
- Protection of key enterprise assets including critical operations, property, people, equipment and intellectual property
- Strengthening reputation and brand through both the application and communication of resilience
- Increased regulatory and governance compliance
- Increased productivity and innovation through integrated and streamlined processes and more effective internal communications

These benefits are documented in the results of the following case studies. While none of the case studies includes maritime or port examples, they do show the importance of resilience and the relevance to transportation and the supply chain:

- Significant return on investment—a major property and casualty insurance firm conducted a study in which it found that those policyholders who fully implemented the preparedness recommendations had on average 75 to 85 percent lower dollar losses than those policyholders who did not implement such measures. Regarding the cost of physical improvements and preparedness, the research indicated a significant return on investment. In the case of Hurricane Katrina, across 476 locations with a total of $42 billion in insured property exposed to the hurricane's impact, the insurance company's clients collectively spent $2.3 million to prevent a projected $480 million in loss, with cost of improvements averaging only $7400 per facility. That equals a 208:1 return on investment. For every $1 spent on targeted preparedness measures, $208 in resources was saved in one single event.[7]

- Increased efficiencies—a Stanford University study, based on inputs from 11 manufacturers and 3 logistics service providers (LSPs), clearly demonstrated that investments in supply-chain security can provide business value. Some of the more significant benefits participating manufacturers reported included the following:
  - Improved product safety (e.g., 38 percent reduction in theft/loss/pilferage, 37 percent reduction in tampering)
  - Improved inventory management (e.g., 14 percent reduction in excess inventory, 12 percent increase in reported on-time delivery)
  - Improved supply-chain visibility (e.g., 50 percent increase in access to supply-chain data, 30 percent increase in timeliness of shipping information)
  - Improved product handling (e.g., 43 percent increase in automated handling of goods)
  - Process improvements (e.g., 30 percent reduction in process deviations)
  - More efficient customs-clearance process (e.g., 49 percent reduction in cargo delays, 48 percent reduction in cargo inspections/examinations)
  - Speed improvements (e.g., 29 percent reduction in transit time, 28 percent reduction in delivery-time window)

FIGURE 6.4

Logistics service providers.

- Resilience (e.g., close to 30 percent reduction in problem identification time, response time to problems, and problem resolution time)
- Higher customer satisfaction (e.g., 26 percent reduction in customer attrition and 20 percent increase in number of new customers)[8]

Resilience and Maritime Security

Resilience Guidance

There are several international standards that are relevant to the issue of developing resilient transportation systems. These include (but are not limited to) ANSI/ASIS Organizational Resilience Standard SPC.1-2009, British Standard 25777:2008 Information and Communications Technology Continuity Management, and the pending International Standards Organization/PAS 22399:2007 Societal Security Guideline for Incident Preparedness and Operational Continuity Management. ISO/PAS 22399:2007 will provide an integrated approach to the management of risk that will include asset protection, security risk management, preparedness, crisis management, emergency management, business-continuity management, recovery management, and disaster management.[9] This sort of construct provides a framework for focusing on resiliency as a primary objective of transportation security regimes and processes.

Integrating Security into Resilience

Resilience is arguably the primary objective for instituting security measures in a transportation system. By including a consideration of resilience in developing and implementing a security regime, measures will likely be adopted that are flexible and agile in responding to potential threats to the system and assuring the system's continued operation even under adverse conditions. According to ASIS SPC.1-2009, the concept of resilience can be analyzed into the following components:

- Prevention and deterrence
- Mitigation
- Emergency response
- Continuity
- Recovery

International Standards Organization/PAS 22399:2007 Societal Security Guideline for Incident Preparedness and Operational Continuity Management identifies the same essential components of resilience but adds the need for a risk assessment to be conducted prior to the "prevention and deterrence" phase. It is important to note that this document is a Publically Available Specification and will be finalized as a Standard in 2012.

The US Department of Homeland Security's (DHS) Risk Lexicon provides an extended definition of resilience as "ability of systems, infrastructures, government, business, communities, and individuals to resist, tolerate, absorb, recover from, prepare for, or adapt to an adverse occurrence that causes harm, destruction, or loss."[10]

Since the DHS definition does not include either a risk-based approach to resilience planning or prevention measures in its definition, it is not complete for the purposes of a fully integrated risk approach to resilience. Additionally, while the ANSI/ASIS Organizational Resilience Standard (SPC.1-2009) does include a prevention component, it does not emphasize the need for a risk-based approach. Therefore, the most complete approach for full-spectrum resilience that can include security measures embedded throughout is the International Standards Organization/PAS 22399:2007 Societal Security Guideline for Incident Preparedness and Operational Continuity Management, which is expected to be released as a fully accepted Standard in 2012.

The Elements of Resilience

The elements of resilience can be analyzed in many different ways. In essence, they all include the following components:

- Detection
- Prevention
- Response
- Recovery

The two elements that are interwoven into all of the other components are risk management and continuity of operations.

FIGURE 6.5

Full-spectrum resilience.

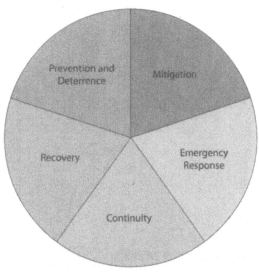

These components are essential in making a security approach truly integrated and a system resilient.

Detection

Detection is the ability to identify the existence of a threat. This may happen at the strategic, operational, or tactical levels. Effective detection includes accurately identifying the threat so an appropriate threat assessment and associated risk assessment can occur. Detection can occur several ways. These may include intelligence information being developed or provided, violations (or attempted violations) of security perimeters or access-control measures, or through the process of screening persons for entry into a facility, other enterprise, or nation.

Prevention

Prevention is the process of creating a nonpermissive environment for the threats to be able to act. This can be done in several ways, including the development of effective security policies, procedures,

---

FIGURE 6.6

Intelligence report.

FIGURE 6.7

Perimeter security.

---

and infrastructure. As deterrence is an element of prevention, physical security is not the only requirement. Creating a security environment that makes it difficult for an adversary to achieve its objectives, may include such measures as physical security, robust response and recovery capabilities, and strong detection capabilities. As will be discussed in a future chapter, the ability to create an environment that allows for the operation of the protected enterprise while deterring an adversary's objectives is optimal in enhancing resilience.

Response

Response is the ability to react to threats quickly and effectively in order to minimize any disruptions to the operations or functions of the protected enterprise or agency. In security terms, responding to a threat or actual event can take the form of increasing security measures, deploying specialized teams, or conducting a rapid assessment of the type of risk that the potential threat poses. Response activities can be designed to mitigate a threat actor's activities by intercepting the threat actor, but can also have a preventive effect by deterring additional activity by a terrorist or criminal adversary. An example of a response that had a preventive effect was the requirement for all aircraft in or headed to US airspace to land immediately after the attacks of September 11, 2001. This response action served to prevent any other potential attacks while the ongoing threats and risks were assessed. The system, however, was arguably less resilient than desired because it required the grounding of all aircraft to ensure that there were no remaining threats.

Recovery

Recovery is an important component of the resilience spectrum, as it is a key determinant of the ability of an enterprise to ensure the continuity of its operations, which is the over-arching objective of

FIGURE 6.8

The new World Trade Center under construction.

a security program. The more quickly an organization can recover from any potential harm if all other measures are fully or partially unsuccessful is a primary indicator of the validity of a security risk assessment and a prioritization of security risks. If an enterprise is harmed but its vital operations remain intact due to an effective emphasis on the protection of critical operations, then the security risk-management process can be assessed as successful.

Continuity

One of the two interwoven components of resilience that apply to all of the above elements is continuity. Continuity is the ability of an enterprise to continue its key functions under varying levels of risk. This is the key determinant of what needs to be protected and how it needs to be protected. The determination of what constitutes continuity may differ depending on the type of enterprise being protected. Continuity for a multinational business may mean the continued ability to produce and ship goods, while continuity for a school may be to continue to hold classes and transport students and teachers safely.

Risk Management

A risk-based approach is also interwoven throughout the aforementioned elements of resilience. Every step requires that there be an assessment made of the criticality of the enterprise, the actual threat, and the level of vulnerability to the essential functions of the enterprise. The details of a risk-based approach to maritime security will be the focus of a future chapter.

The Medical Comparison

In order to establish an integrated security approach within the above elements of system resilience vs. system security alone, it is useful to use the example of a medical model of care as a way of understanding the integration of security into a resilience approach. Ignoring any current political arguments regarding the state of healthcare in the United States, healthcare can be understood to have the objective of focusing on an individual's ability to prevent, withstand, and recover from potential medical threats such as disease and treating other risks as necessary. Therefore, the medical healthcare process for individuals can be divided into elements similar to the elements of resilience:

- Prevention
- Detection
- Diagnosis
- Response
- Recovery

Diagnosis can, in the resilience and security spectrum, be included within the detection element, where threat assessments are produced, and throughout the spectrum, where risk assessments perform the diagnostic function. The graphic below shows the relationship between the medical case model and the resilience spectrum.

Enabling Resilience

By weaving the key elements of security into the resilience spectrum of activities, each element is linked with each other and is complementary. This approach can minimize duplication of efforts, working at cross-purposes, and demonstrate the security profession's inherent value in enterprise resilience.

FIGURE 6.9

Comparison between medical case management and security integrated into resilience.

Medical Case Model	Medical Case Model Activities	Resilience Spectrum	Resilience Spectrum Security Activities
Prevention	Patient awareness, exercise, and diet recommendations	Prevention	Security measures, perimeter security, awareness
Detection	Physical examinations, health screenings	Detection	Threat assessments, access control, screening
Diagnosis	Lab testing, research, examinations	N/A; embedded in detection and response	Case forensics, investigations, assessments
Treatment	Medicine, therapy, surgery, monitoring	Response	Risk treatment, emergency response, crisis management
Recovery	Physical therapy, rest, additional treatment	Recovery	Business-continuity planning, resumption of operations

FIGURE 6.10

A highway-port connection in New Jersey. *Photo credit: US National Oceanographic and Atmospheric Administration.*

A full-spectrum approach to security and resilience builds in efficiencies, as response plans and measures can be used for other events, including security incidents. As will be shown in a future chapter, this approach will lower the costs of security and make a compelling business case.

A key challenge in adopting a full-spectrum, resilience-focused approach to maritime and port security is the determination of the scope of security measures required. Maritime security, due to its nature, touches other infrastructure sectors. Therefore, it is important that as intermodal or intersectoral connections are identified, the implications of those connections to the resilience-based approach are properly considered. An example is the need to factor in the implications of the intermodal supply chain in enterprises where it is a critical requirement for continued operation. In cases like a supply chain, a resilience approach could focus on the integrity of the supporting supply system rather than just the physical security measures in place. A proposed regulatory framework to address this issue will be presented in the next chapter.

REFERENCES

1. American Society of Civil Engineers, *Guiding Principles for the Nation's Critical Infrastructure,* Reston, VA: 2009, p. 21.
2. Deloitte Research. "Disarming the Value Killers," Deloitte, February 2006.
3. "The Value of Resilience," Council on Competitiveness, Council on Competitiveness, 2006.
4. "Can Your Business Survive A Natural Disaster?," Alfa Insurance, 2007.
5. "The Effect of Supply Chain Disruptions on Long-term Shareholder Value, Profitability, and Share Price Volatility," Kevin Hendricks and Vinod Singhal, The Logistics Institute, June 2005.
6. New York University: International Center for Enterprise Resiliency, *The Business Case for Enterprise Resilience: Significant Risks, Substantial Rewards.*
7. "Crediting Preparedness," William G. Raisch, Director, and Matt Statler, Ph.D., Associate Director, International Center for Enterprise Preparedness, New York University, 8/2/2006.
8. "Innovators in Supply Chain Security: Better Security Drives Business Value," Barchi Peleg-Gillai, Gauri Bhat, and Lesley Sept, Stanford University, July 2006.
9. Siegel, Marc. *Societal Security Management System Standards.* ASIS International, 2008.
10. Risk Steering Committee. DHD Risk Lexicon, 2010 Edition, p. 26.

CHAPTER 7

Standards and Regulations

INTRODUCTION

As reviewed previously, the only globally mandated standard, code, or regulation pertaining to maritime security is the International Ship and Port Security (ISPS) Code. The ISPS Code, while providing an effective initial baseline set of requirements for ships and ports, needs to be improved, based on almost a decade of use and associated lessons learned. However, it is important to emphasize that regulations, codes, and standards are tools to be used in improving port and maritime security and are not in themselves objectives. The full implementation of any international requirements is a process that is never fully completed and requires regular review and modification based on changes to operations or risk. Further, any consideration of additional requirements should be analyzed with the express intent of ensuring that the standards do not unduly inhibit trade and commerce.

REVIEW OF THE ISPS CODE

Prior to discussing possible improvements to international maritime security requirements, it is useful to review the basic tenents of the ISPS Code and its weaknesses, which were identified in more detail in Chapter 5.

The ISPS Code

The ISPS Code is a reasonably effective initial step in establishing low-level baseline security in global shipping. Of particular note is the design of the ISPS Code to focus on the desired security outcomes without being overly prescriptive in the manner in which they are realized. This is particularly effective because of the drastic differences in the size, technological development, and resources available to ports, administrations, and shipping companies around the world. As a result, the ISPS code is divided into two sections, Parts A and B. Part A consists of the requirements of the ISPS Code that need to be implemented in order to be compliant. Part B includes more specificity on the requirements laid out on Part A but is not mandatory. Further, while the Code requires national governments to oversee its implementation, it does not prescribe specific roles for government and the private sector since those roles will vary from port to port or nation to nation. However, the ISPS Code has several inherent weaknesses, including the following:

- The ISPS Code focuses primarily on external threats and does not address supply-chain security issues as a primary issue. Its focus on external threats creates a vulnerability regarding the potential for internal conspirators to initiate criminal activity or terrorist acts from inside the fence line of an ISPS-compliant facility. The ISPS Code addresses security in the context of the port or ship as a potential target vs. a conduit for illicit material or people.

- The ISPS Code relies on self-regulation. It requires national governments to oversee its implementation and to approve security plans and training, facilities, and ships or companies are expected to identify their own critical operations and to carry out their own threat and vulnerability assessments using in-house non-security specialists with limited training.

- There is no public information regarding how potential adversaries rate the effectiveness of the ISPS Code. While there may have been an analysis of its effectiveness by security or intelligence agencies, it has definitely not been shared with corporate security officers or others responsible for the implementation of the Code.

- The ISPS Code, written to promote self-regulation, has an inherent assumption that all parties involved in day-to-day port business or operations are legitimate actors. The Code does not provide detailed requirements for the conduct of background checks for employees or authorizations for visitors, thereby creating a significant variance in access to port areas among countries.

- Regulations compel compliance but do not necessarily effectively build security into daily operations or the organizational/corporate culture. The ISPS Code has been adopted by most countries in the world that have ports or shipping fleets. In most cases, the Code is adopted outright and is made regulatory through implementing legislation. Some countries implement national port-security legislation that exceeds the requirements of the Code.

Some of these weaknesses are more easily solved by providing additional, well-constructed international requirements, while others are the result of cultural and institutional considerations that are less easily fixed or improvedin this manner.

ISPS CODE 2.0

Since its implementation in 2004, the ISPS Code has been adopted throughout most of the world, and sufficient time has transpired to determine its efficacy. During that same period, there has been increased focus on the security of the global supply chain as well as the ports and ships that handle most of the world's international commerce. Therefore, in order to reflect the concern about the security of the supply chain, any revisions to the ISPS Code should include security requirements that address supply-chain security in addition to the existing requirements. The inclusion of supply-chain security requirements also addresses one of the previously identified weaknesses of the ISPS Code, that it is not effective against threats that seek to exploit ports and shipping as conduits for illegal activity or material. The basic elements of an improved ISPS Code would consist of:

- Use ISO 28000 with modifications as the foundation for a new Code

- Add a ship security component to the existing requirements in ISO 28000 and 20858 to ensure the current ISPS Code requirements remain in place
- Ensure that the port and ship security requirements that are incorporated into the new Code include both Parts A and B of the current ISPS Code as mandatory
- Identify the authorized economic operator (AEO) concept within the new Code as a model for a risk-based approach to supply security.
- Include resilience and information security standards

Use ISO 28000 as the Foundation for a new ISPS Code

International Organization for Standardization: ISO 28000 (International Standard 28000:2007 Specification for Security Management Systems for the Supply Chain) is particularly beneficial to maritime security because it provides general guidance on the implementation of supply-chain security measures while also including a port-security standard in ISO 20858:2007 (Ships and Marine Technology Maritime Port Facility Security Assessments and Security Plan Development) that is fully compatible with the ISPS Code, including Parts A and B for ports. However, ISO 20858 does not include a standard for ship security, which results in a remaining gap that would need to be added to ISO 28000 to make it fully compliant with the current ISPS Code.

Another subsidiary standard to ISO 28000 is ISO 28001:2007 (Security Management Systems for the Supply Chain Best Practices for Implementing Supply Chain Security, Assessments and Plans Requirements and Guidance). ISO 28001 amplifies the general approach of ISO 28000 by providing more detailed guidance in the implementation of supply-chain security measures. Of significant note, ISO 28001 specifically endorses the EU's authorized economic operator (AEO) concept, which focuses on assessing the risk of companies and other organizations that are participating in the supply chain and providing varying levels of controls on those actors within the supply chain based on their assessed risk. The US government has also reached an agreement with the EU in which the AEO program will be deemed fully compatible with the requirements of the Customs-Trade Partnership Against Terrorism (C-TPAT).

Additionally, the incorporation of existing resilience standards to create linkages with existing response and contingency plans will be important in ensuring that existing non-security response plans are included, thereby minimizing the need to duplicate planning efforts. There are several accepted resilience-oriented standards that could be incorporated in the new Code. The primary examples are ASIS SPC.1-2009, Organizational Resilience, which is approved by the American National Standards Institute (Organizational Resilience: Security, Preparedness, and Continuity Management Systems Requirements with Guidance for Use) and BS 25999-1:2006 (Business Continuity Management Part 1: Code of Practice). A new code should be outcomes-based rather than prescriptive. Therefore, the development of outcomes-based standards and regulations, which mandate required security outcomes while allowing the specific methods of achieving those outcomes to be tailored to the needs of stakeholders, is effective for several reasons:

- They allow for desired security outcomes, as articulated by strategies and plans, to drive the procurement of technology and infrastructure through the development of system requirements.

- Outcomes-based approaches do not require significant construction projects unless the required security outcome cannot be met through modifications to existing infrastructure.

- Outcomes-based standards focus on the performance of core security missions rather than simple conformance to performance or technology standards.

Outcomes-based standards are derived from policy decisions, strategies, and plans that reflect the border-security priorities of the national and agency leadership. Once outcomes-based performance standards are identified, the system requirements for technology and equipment are then developed, which, in turn drives the procurement process.

Finally, any new maritime security code should incorporate the principles of ISO 27000 (Information Security Management Systems) in order to ensure the integrity of the overall management system.

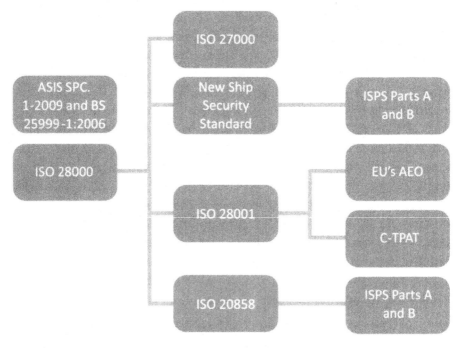

FIGURE 7.1

ISPS 2.0.

_____

Considerations

In implementing a new international maritime security code, there are a number of issues that need to be considered to ensure proper and effective adoption. These issues can be placed into two key categories: issues that may affect acceptance and issues that may affect implementation.

Acceptance Issues

Issues that may affect acceptance of a revised code by stakeholders include:

- Recognition that the original ISPS Code served as a starting point for securing the maritime domain rather than a definitive and final end result
- Consultation with a broad array of stakeholders

- A commitment to ensure that any additional or new requirements are not excessively burdensome and are uniformly applied to ensure that economic competitiveness is not adversely affected
- Where possible, making a clear business case

Recognizing the Need to Revisit the ISPS Code

While most commercial port and vessel operators are likely content to continue abiding by the ISPS Code because the primary investment in security infrastructure, planning, and training has already occurred, there is likely to be significant objection to additional or revised requirements unless a compelling justification can be made. A potentially effective approach includes an assessment of the individual regulatory requirements that the private sector is committed to or required to comply with, which increasingly includes supply-chain security measures and, in many countries, a de facto requirement to comply with the ISPS Code's Part B as it is enshrined in national port and vessel security legislation. An example of this is the US Maritime and Transportation Security Act (MTSA) which is more stringent than the ISPS Code. Therefore, an argument can be constructed that the consolidation of disparate security programs and requirements into one code may provide a clearer roadmap for the private sector.

Consultation

In order for a major revision of the ISPS Code to be accepted, a wide variety of stakeholders would have to be engaged and encouraged to provide their inputs to any proposed changes. Prior to getting their input, they will have to be convinced of the necessity and utility of revising the code. Potential stakeholders may include:

- Port states
- Flag states
- Port operators
- The insurance industry
- Ship operators
- International organizations including IMO, ISO, and WCO
- Large shippers
- Representatives of industries reliant on shipping

FIGURE 7.2

IMO plenary session.

Equitable Application of a New Code

A common criticism of the current ISPS Code is the perception of uneven implementation among signatory nations. Some countries, such as the United States, United Kingdom, Australia, and the European Union, have developed mature and detailed implementation requirements in their respective enabling legislation. For these countries, both Part A and the more prescriptive Part B have been included in their regulations. Since countries are only required to adhere to Part A to be considered in compliance, other countries have adopted a "lighter" touch, which generally makes compliance easier and less costly. The lack of an even approach has led some critics to be concerned that the ISPS Code is not uniformly applied and may be more burdensome to some vessel and port operators than others. Therefore, any revisions to international maritime security codes must be constructed in a way to be uniformly applied while retaining the outcomes-based nature of the existing code to allow for tailored implementation in a variety of ports and among a diverse population of ships. This approach will likely contribute to the acceptance of new requirements if they are perceived as being fair while retaining the flexibility to meet prescribed outcomes with locally appropriate solutions.

The Business Case

In previous chapters, examples of the lack of security or resilience planning clearly demonstrated the potentially dire results for businesses. Future chapters will explore the compelling need to establish a business case for security and resilience as well as potential approaches to developing it.

Implementation Issues

Issues that may affect implementation may include:

- Private-sector concerns over costs and burdens of implementation
- Funding for security improvements
- Consistency of new requirements with existing national or international regulations
- Bureaucratic infighting and workload challenges

Private Sector Burden

The private sector will inevitably be required to expend funds in order to implement any new Code. However, by including existing Codes and widely accepted standards, the level of required spending should be manageable. Clearly, a detailed study would need to be completed during the consideration phase of an expanded Code in order to accurately assess the potential industry burden. Currently, the more reputable private-sector elements already abide by much of what would be included in a new Code. Therefore, theist adoption would serve to put those companies that have invested in security and resilience beyond the current minimum requirements in an advantageous position by having already made significant progress and efforts.

For smaller private-sector entities that would be particularly burdened, the Code could be prepared in such a way as to allow for the pooling of management systems in order to reduce costs where feasible.

Funding Challenges

In some cases, the implementation of new requirements will not require the expenditure of significant amounts of additional funds, as some ports and shipping companies already comply with both the enhanced port and ship security regulations as well as the additional supply-chain security requirements. In other cases, especially in the developing world, where the implementation of the ISPS Code has focused primarily on Part A, there may be a significant challenge to pay for the implementation of additional requirements. In these cases, all governmental grant programs and relevant foreign-aid packages should be tied to the outcome of fulfilling the requirements

FIGURE 7.3

International security aid provided by the US Coast Guard in Iraq. *Photo credit: US Navy.*

of the new Code. As has been described previously, the current US port-security grant program does not effectively target monies to measurable outcomes, and international aid programs often do not have mechanisms in place to ensure that the aid provided is applied correctly.[1] Therefore, grant and aid programs will have to attach more targeted performance goals and verification procedures to ensure that the expenditure of funds are directed most effectively. In short, government budget and grant considerations should be tied to demonstrated effectiveness and adherence to resilience- and security-based standards.

Regulatory Compatibility

Any new or revised international maritime security code should be crafted to ensure that it is compatible with other related codes or requirements. For example, the current ISPS Code is a component of IMO's broader Safety of Life at Sea (SOLAS) Convention, which has a broader remit than security and includes well-established maritime safety requirements. The SOLAS Convention is globally accepted and recognized as the world-wide standard in maritime safety. Virtually all national port-state control programs are based on the requirements outlined in the SOLAS Convention, and the ability of port-state control programs to work are dependent on the framework provided by it.

There are several key supply-chain security programs or standards that are internationally accepted in addition to ISO 28000. The World Customs Organization's SAFE Framework provides guidance that is consistent with the broader ISO 28000 standards, as does the European Union's AEC guidance and the US Customs-Trade Partnership Against Terrorism (C-TPAT). Any new international security standards need to be developed so that they are compatible with the currently accepted, nonmandatory supply-chain security standards.

Bureaucratic Challenges

A major revision of the ISPS Code to include a significant supply-chain security element may result in competition and infighting among bureaucratic organizations or agencies charged with implementing the new code as well as concerns about increased workload and responsibilities that may not have been reflected in agency personnel and budgetary requirements.

If a new maritime and port security code that includes, and indeed emphasizes, supply-chain security were to be adopted internationally, a major concern of many governments throughout the world would likely be how to internally organize in order to carry out the requirements. The proposed code would blend elements of responsibilities that are typically found in national customs services, port-state control authorities, border-security agencies, and

---

FIGURE 7.4

US Coast Guard K-9 team. *Photo credit: US Coast Guard.*

coast guards. Therefore, there is likely to be concern among governments and agency chiefs regarding how government agencies can be
realigned or their activities coordinated to meet the requirements of
the new code.

For example, in many countries there may be a combined
border-security agency that includes customs enforcement or a traditional construct of a customs service and an immigration service.
Additionally, the agency responsible for the implementation and
enforcement of the current ISPS Code is likely not to be a border
agency but may be a police or coast-guard organization or a nonop-
erational government agency such as the Office of Transport Security
within the Australian Department of Infrastructure and Transport
or the Department of Transport in the United Kingdom. Therefore,
implementation of a broader code may require significant stakeholder management in order to overcome bureaucratic insecurities
over authority and jurisdiction.

Much of the governmental concern in implementing the new
code will lie in the legislative changes that may be required to carry
out the requirements of the new code and any changes in governmental agency authorities or missions to reflect the changes. Further,
government bureaucrats who are invested in particular agencies may
feel threatened by potential changes in their authorities, responsibilities, agency cultures, and resource availability (especially budgetary)
and may therefore, be resistant.

Other Implementation Considerations

The development of a new code could bring the opportunity to make
improvements to the overall approach to maritime security by adopting a more genuine risk-based approach and by implementing a more
dynamic process for validating and revising any plans and procedures
included in a new code.

Embedding a Risk-Based Approach

While the specifics of developing a truly risk-based approach to
maritime and port security will be developed in a future chapter,
the revision of the international port-security code will provide an
opportunity to develop more effective and sophisticated risk treatments as well as a process for determining the risk appetite of the key
involved parties.

Continuous Improvement

While the ISPS Code requires periodic reviews of the associated plans and procedures as well as triggers for increasing security measures and conditions, a revised code based on ISO 28000 will allow the inclusion of a quality-based, continuous-improvement process that is not solely tied to regulatory requirements. The adoption of a plan-do-check-act (PDCA) approach to the design and implementation of the plans and procedures included in a new code will likely allow for adjustments based not only on governmental requirements and changes in threat levels but also on changes in business operations and requirements. A PDCA approach allows for continuous improvement and evaluation of efficiency. Further, the use of a PDCA approach may include a three-tiered audit system that could ensure greater compliance with the code than is currently guaranteed.

PDCA, found in many quality-management systems and common among internationally recognized standards, consists of an ongoing, repetitive cycle, whereby a standard, code, or practice is continually evaluated for its appropriateness and application and adjustments are made based on emerging or changing requirements. It also facilitates compliance by encouraging regular evaluation of the effectiveness and efficiency of the standard's implementation.

Typically, international standards have three levels of auditing. These include internal auditors drawn from within the organization responsible for implementing the standard, who ensure that basic compliance measures are adhered to; external auditors, who are

FIGURE 7.5

Plan-do-check-act.

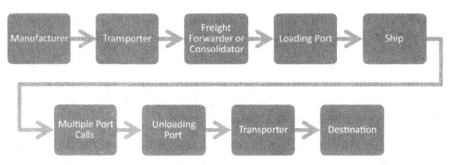

FIGURE 7.6

Supply-chain security's reach.

---

brought in by the responsible organization but have a potentially more neutral approach, as they are not responsible for the implementation of the standard; and the third-party auditors, who may represent an external certifying body. Currently, the ISPS Code utilizes a similar system where the implementing organization is responsible for conducting internal audits and the designated authority of a country is responsible for conducting the "third-party" audit, sometimes through a recognized security organization acting on its behalf.

Encouraging a Management System Approach

The inclusion of standards that are configured to comply with the International Organisation of Standards (ISO) requires the establishment of a management system for the implementation and maintenance of the elements of the PDCA cycle. Fortunately, each organization only requires one management system, as the elements of other ISO codes can be added, thus building in organizational efficiencies.

Notional Contents and Structure of a New Code

Building on the ISPS Code, a new, comprehensive maritime security code would incorporate the requirements of the ISPS Code with supply-chain security standards to ensure a fuller treatment of the security issues surrounding the maritime domain. Additionally, the inclusion of supply-chain security measures will provide excellent security linkages to other modes of transportation and infrastructure. Currently, the ISPS Code's effectiveness is not only limited by its minimal consideration of the potential for ports to be exploited as conduits for the movement of illicit material or people but also by the limits of its jurisdiction to the boundaries of a specified port area.

A more comprehensive, international code would also include a more mature approach to risk assessment and the treatment of identified risks that is in accordance with ISO 31000. This will be addressed in more detail in a future chapter.

Finally, the inclusion of several existing standards and codes into a single code will likely result in some economies of scale by the private sector, which will either be required or strongly encouraged to adopt or participate in programs derived from the standards cited below.

The New Code

A proposed code that incorporates supply-chain security will have elements from the following existing codes and standards to provide a more holistic approach to supply chain security.

Each standard and code serves to enhance the new code's effectiveness while not requiring redundant measures on the part of ports or ship operators, thus minimizing the impact of implementation. Specific contributions of each element of the code are:

- ISO 28000 (Series)—provides overarching supply-chain security management system guidance
- ISPS Code (Parts A and B)—provides specific port and facility security requirements
- European Union's Authorized Economic Operator Guidelines—provides a mechanism for assessing the risks posed by participants in the supply chain and ensuring recommended controls; compatible with the US C-TPAT program
- World Customs Organization SAFE Framework—provides specific customs measures to enhance security
- ANSI Resilience Standard—provides a management-system approach to preparedness, planning, and response
- ISO 31000—provides the overarching standard for the code's approach to risk management; requires some additional modifications regarding the assessment of

threat and risk appetite, which will be covered in a future chapter

- ISO 27000 (Series)—provides the standard for information security to be used throughout the code's development and implementation

The establishment of a comprehensive maritime security code that includes the current ISPS Code but also a broader approach to supply-chain security based on internationally recognized, auditable standards with an associated management system will serve to fill the gaps and weaknesses identified in the ISPS Code and the current state of its implementation.

REFERENCE

1. The author has direct experience in developing countries with examples of foreign security aid that obviously did not reach its intended targets and was redirected. However, without physically assessing the level of aid reaching ports and security forces, officials in many national capitals likely believe that the programs are working.

CHAPTER 8

Assessing and Managing Risk

INTRODUCTION

Effective management of risk is directed towards the management of potential opportunities for improvement (good risk) and the reduction of adverse effects (bad risk). The use of risk-management techniques is integral to business planning through the use of a process that will assess the costs and benefits of options open to identify and reduce areas of risk. The use of a structured process ensures that it is transparent and repeatable, which also contributes to a model or risk program that can be continually monitored and updated without undertaking the whole exercise constantly. It should also be noted that while the risk-management process seeks to measure the level of risk and the objective outcome, the nature of the assessment is in many ways subjective, as it is based upon the interpretation of information. The more rigorous the process, the more accurate the assessment—this is the answer to establishing a reliable risk model.

The development of a risk-assessment methodology ensures that structured processes are in place to produce a comprehensive assessment that can be easily followed and then repeated at regular intervals throughout the life cycle of the project, providing a continuous risk picture set against a common level of understanding of the threat and operating environment.

The International Standards Organization Standard 31000 on Risk Management Principles and Guidelines (ISO Standard 31000), outlined in Figure 8.1, is a useful starting point for developing a rigorous process to identify, analyze, and manage maritime security risk. In order to be most effective, however, the standard will require some additional focus on the areas of threat assessment and the assessment of risk appetite or tolerance, both of which are key elements in the development of a realistic, rigorous, and accurate risk assessment. The initial threat-focused stages of the methodology are designed to provide an understanding of the threat environment in which the maritime domain is situated. By using an "intelligence-led" approach, the methodology enables changes in the risk environment to be anticipated. As a result, regular monitoring of this environment is a critical part of the risk-management process.

ISO 31000

The ISO 31000 risk-management process consists of six core components. Five of the components are sequential, while the sixth (monitor and review) is an ongoing component that is included throughout the process and ensures currency and effectiveness. Further, an embedded component is a communicate and consult function, which is critical to ensuring the acceptance and the legitimacy of any

---

FIGURE 8.1

ISO 31000 risk-management process.

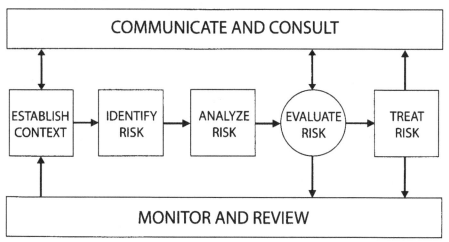

assessments and proposed treatments. This is particularly important in complex risk-management problems involving a large number or a varied group of stakeholders:

- Establish the risk-management context
- Identify risk scenarios
- Analyze risks, including consequences and likelihood of identified scenarios
- Evaluate risks by ranking risk scenarios relative to one another;
- Treat risks by developing and implementing cost-effective strategies and action plans for increasing potential benefits and reducing potential costs
- Monitor and review risk and the effectiveness of treatment measures to ensure that changing circumstances do not alter priorities

Integrating this risk-management process into all maritime security operations provides the security manager with an understanding of the security risks present within his or her areas of responsibility and associated maritime and intermodal transportation modes, which in turn enables key decision-makers to make targeted planning, budgeting, and policy-making decisions. The risk-management process can be scaled for applicability at tactical and strategic levels, thereby allowing for in-depth understanding of security risk at individual maritime facilities, comparison of risk across facilities or geographic areas, and consideration of broader security challenges and potential risk-mitigation investment strategies in national or international sectors.

Despite its overall utility and wide acceptance, ISO 31000 does not include specific guidance on how to determine risk appetite or tolerance, which is essential in the prioritization of risks and the development of a risk register. Further, as n oted in previous chapters, any truly effective security risk assessment methodology require a firm grounding in understanding threats and threat assessments. Finally, a risk register needs to consider potential security threats or hazards but cannot easily do both because of the differing methodologies used. Therefore, an all-hazards risk assessment is by its nature less likely to comport with the requirements for an in-depth threat assessment that is key to effectively conducting a security risk assessment. If necessary or desirable, a separate assessment using much of the same methodology could be carried out for nonsecurity hazards. The results of both assessments could then be qualitatively compared to determine an overarching risk-management approach without combing them into one risk register.

Risk Terminology

The following are terms and definitions that are used in the assessment and management of risk, derived from ISO 31000.

Risk

Risk is the effect of uncertainty on the objectives of an organization and is often expressed in terms of the combination of the consequences of an incident and their likelihood. As noted in previous chapters, this definition is significantly different form the definition used by the US Department of Homeland Security (DHS), which defines risk as being of an inherently negative nature. The DHS definition is limiting because the options for risk treatment, thereby only be mitigated. A neutral definition of risk, as found in ISO 31000, allows for a more nuanced approach to risk treatment.

Risk Management

Risk Management is the coordinated activity to direct and control an organization with regard to risk.

Risk Assessment

Risk Assessment is the overall process of risk identification, risk analysis, and risk evaluation.

Risk Analysis

Risk analysis is the systematic process to comprehend the nature of risk and to deduce its level.

Risk Appetite or Tolerance

Risk appetite is the amount and type of risk an organization is willing to accept.

Other Definitions

The following definitions are not derived from ISO 31000 but provide a differentiation between threats and hazards in order to clarify the differences in assessment elements for each.

Threat

Threat is a credible situation with a potential *intentional* event for injury or damage to assets. Threats emanate from terrorist or criminal actors, either individuals or groups. The methodology for conducting a threat assessment will be included as an appendix but generally consists of an analysis of the capability and intent of a terrorist or criminal actor. Threat assessments are critical when conducting security risk assessments. An analytically rigorous threat assessment provides the foundation for the rest of the risk assessment by allowing decision-makers to understand the capability and intent of the adversary. This understanding provides the cornerstone to effective and sophisticated risk management in the security discipline.

Hazards

Hazards are natural or accidental incidents that emanate from *unintentional* human action or acts of nature. Since hazards are unintentional events, they are usually evaluated by analyzing historical data and probability research to identify the potential for a hazard to occur.

Vulnerability

Vulnerability is a weakness of an asset or group of assets that can be exploited by one or more threats.

Likelihood

Likelihood is an estimate of the chance of an incident happening and reflects the combination of threat and vulnerability assessments.

Consequence

Consequence is the outcome of an event or change in circumstances affecting the achievement of objectives or normal operations. When assessing consequence in more complex situations, it is important to identify the levels of consequences to be considered. Consequences that do not directly derive from an event are often described as cascading events and may have longer-reaching effects.

Core Components of Risk

The core components of risk form the step-by-step approach to carrying out a well-planned risk assessment and developing a truly risk-based approach to security.

Establishing the Risk Management Context

This step of the risk-management process involves consideration of the goals and objectives of risk management, including the management and operational decision(s) to be supported by the risk assessment, the goals and objectives these decisions support, and an understanding of the security priorities and constraints of relevant decision-makers and other stakeholders. To ensure transparency of the risk-management process, it is important to clearly document the assumptions that supported the identification of the risk-management context. As requirements change, it may be necessary to revisit these assumptions in the future.

At the outset of a risk-management activity, it is useful to gauge the extent to which decision-makers are prepared to tolerate risk. Understanding decision-makers' risk appetite facilitates the development of strategies for prioritizing and mitigating risk.

FIGURE 8.2

Establishing the risk management context is the fist step in assessing risk.

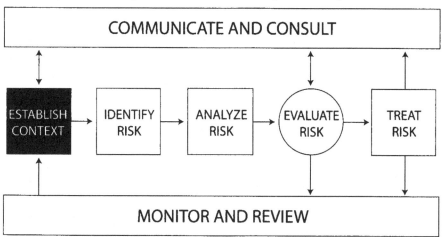

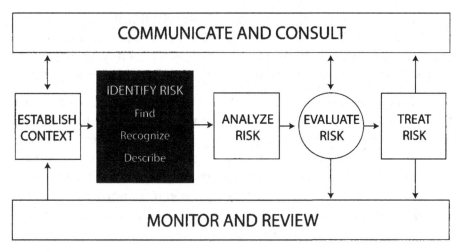

FIGURE 8.3

Identifying risks within the established context provides the foundation for analysis.

Most risk-assessment methodologies either do not address or address in very cursory manner the issue of identifying the risk tolerance of the subject of the exercise. If it is accepted that risk management is really about managing risk, not just mitigating or reducing risk, the determination of risk tolerance is of vital importance. This process needs to be included in risk assessments so the appropriate and tailored treatment measures can be developed to meet the requirements of the protected entity.

Risk tolerance is the amount of risk that can be accepted by a person or entity without the requirement to treat the risk. An established process to include this in risk-management methodologies or approaches will serve to more effectively identify the critical elements of an organization or operation and will likely prevent tendencies to risk aversion.

Identify Risks

Risk identification is the process of finding, recognizing, and describing potential risks. In security risk assessments, the identification of risks typically begins with consideration of what could go wrong, including noting targets or assets to be protected and consequences (loss of life, economic impacts, and reputational impacts) of concern to decision-makers. As part of this step, potential threat actors are identified and the significance of the threat they pose is assessed against the potential target, whether it be a ship, port, facility, or region. A detailed threat-assessment methodology is included in Appendix B.

Scenario development is a key step in the risk-identification process. Scenarios should be worst case-most likely case and contain a realistic depiction of potential threats after they have been assessed. Analysts use historical incidents for scenario ideas during the threat-assessment process, but any indications of shifts in the targeting or capability of threat actors should be included if supported by an analytically rigorous process. Rigorous research by threat and industry experts is necessary to prepare valid scenarios.

A worst case-most likely case scenario takes into account all aspects of risk: threat, vulnerability, and consequence but does not include situations that are atypical and highly unlikely to occur.

An initial step in scenario development is to identify and assess the potential threats that could impact the asset(s) or target(s) of concern. Within each threat category (terrorist attacks, smuggling activities), analysts should brainstorm and define potential scenarios based on the research performed for the threat assessment. As noted in Chapter 4, the term "threat" is usually meant to define the potential source of action or activity that is being protected against. Some definitions, including those by the US Department of Homeland Security, include natural disasters and other catastrophes as threats. This broader definition, while simpler to use, weakens the analytical rigor needed to focus on security threats by mixing in events that are not controlled by humans who are actively trying to circumvent protective measures. Therefore, the threat-assessment process is given less emphasis, as there is a broader swath of potential crises that cannot be prevented, which shifts the focus of activities to response or impact-mitigation planning. This lowered emphasis on accurately assessing security threats by mixing them with safety threats has led to a weakening of the threat component in "all hazard" risk assessments and therefore has drawn focus away from effective and preventive security risk treatments.

For the purposes of this work, the definition of threat will focus on the potential "bad actors" who knowingly and intentionally target the maritime operating environment for harmful exploitation. All other potential impact events or sources will be termed as "hazards." This is important, because in later chapters we will focus on threat assessments, which are performed in a completely different way than hazard assessments.

The process of assessing threats requires a significant amount of analytical rigor and a clear methodology to ensure that threats and their associated scenarios are correctly identified. Further, threat assessments are perishable due to the regularly changing nature of threat groups and their associated targets and tactics and should be reevaluated on a periodic basis unless a significant security event occurs that spurs a more frequent reevaluation.

As noted in previous chapters, is it important that threat assessments be analytically rigorous and that the pitfall of threat advocacy be avoided. Threat advocacy is a situation where, in the absence of a fully justified threat assessment, certain individuals will develop "what if" scenarios that are not fully assessed and may be focused on the desire to justify certain pre-ordained views or desired security outcomes. Brian Jenkins, an internationally recognized terrorism authority at Rand, argues that much of the national debate in the US is centered around threat advocacy as opposed to threat assessment, and that elements of the national security community may have concerns that do not correspond with the adversary's actual strategic and tactical goals.

After identifying and assessing potential threats, the next step in the risk identification process is to build plausible risk scenarios that match attack methods to assets or targets of concern. When developing risk scenarios, it is important to include any reference material or sourcing that supports the plausibility of the scenario. The risk scenarios developed in this stage form the basis for the risk assessment.

Analyze Risks

The risk-analysis process should be designed to be flexible and repeatable and should answer the primary three questions of security risk:

- What can happen?
- How likely is it to happen?
- What are the consequences if it does happen?

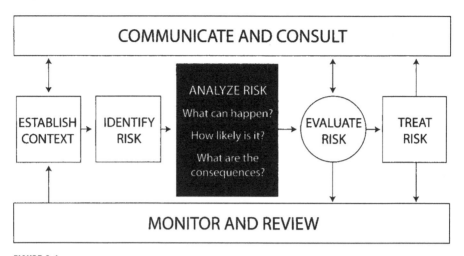

FIGURE 8.4

Analyze risk by asking these three key questions.

Threat and vulnerability together can be considered to determine likelihood, while consequence captures the effects of interest to the appropriate corporate or governmental decision-makers. In this step, the threat scenarios developed previously are applied to the intended target and the likely consequences are identified. Consequences may include fatalities, loss of capability, physical damage, monetary losses, reputational damage, or other consequences as appropriate. In the case of shipping, the issue of cascading effects on other political, economic, and social sectors may need to be included if there are disruptions to the supply chain.

Evaluate Risks

The purpose of risk evaluation is to assist in making decisions, based on the outcomes of risk analysis, about which risks need treatment in order to prioritize implementation. The evaluation of risk scenarios involves the ranking of these scenarios relative to one another to identify those that are of highest risk. It is often useful to develop graphical displays, such as risk registers, risk matrices, or other ranking frameworks for presenting the results of the risk assessment to decision-makers in an easily understood format. The results of risk assessments can be presented for individual facilities or aggregated across facilities and sectors or regions to support the identification of trends and systemic risks to inform strategic policy-making. Regardless of the presentation method selected, the results of the risk assessment and the underlying methodology should be fully documented to ensure the transparency and repeatability of the process.

Risk matrices are useful tools for presenting the findings of risk analyses to decision-makers or other stakeholders. These matrices provide a summary snapshot of the risk analysis findings and generally are accompanied by narrative analysis. An example of a risk register in graphic form is found in Appendix A. The scenarios used to populate the examples are for illustration only.

As noted previously, the increasingly common "all hazards" approach to the development of a risk register is flawed due to the different methodologies used to assess threats and hazards. If an all-hazards approach is desirable for an assessment, two discrete risk registers should be developed. When both risk registers are developed, the results can be considered together through consultation among key stakeholders. During this process, the key risks from both registers can be evaluated from a strategic perspective and prioritized. This approach is most likely to be effective where an enterprise risk approach is adopted that comprehensively addresses all risks that may be pertinent to the assessed entity, whether it is a government, company, building, or any other entity. In enterprise risk management,

FIGURE 8.5

Evaluate risk by ranking scenarios and identifying the highest risk.

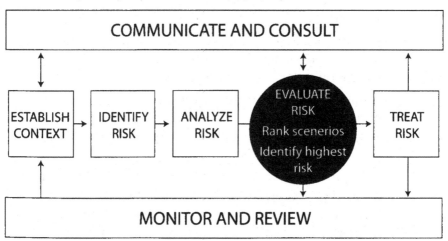

risks that need to be considered in addition to security or resilience may include:

- Reputational risk
- Financial risk
- Legal risk
- Strategic risk
- Operational risk

In enterprise risk management, the consequences and threats or hazards to one area may have an effect on other areas. Therefore, it is within the context of enterprise risk management that the all-hazard approach should be pursued. However, separate risk registers must be developed for threat-based risks and hazard-based risks.

Treat Risks

In this phase, appropriate responses and risk-mitigation strategies are developed to treat identified high-priority risks. Considerations include the risk tolerance of stakeholders, the potential cascading effects of risk-treatment decisions, cost effectiveness, and timeliness.

The relative ranking of risks enables decision-makers to prioritize risks for the potential application of treatments. Risk treatment involves selecting one or more options for addressing risks in accordance with the agreed-upon risk-tolerance analysis. In treating risks, decision-makers can consider a number of options:

- Accepting the risk by not implementing any countermeasures.
- Avoiding the risk by discontinuing the activity that presents a risk or instituting measures that mitigate threat, vulnerability, or consequence.
- Reducing risk by putting in place risk-mitigation measures—this is the most common approach to addressing risk when a fully developed risk-management program does not exist.
- Recognizing that the identified risk is too significant to be avoided or accepted but cannot be mitigated, therefore, transferring to another entity is a viable option—the insurance industry is the most commonly cited entity in the risk transfer business.

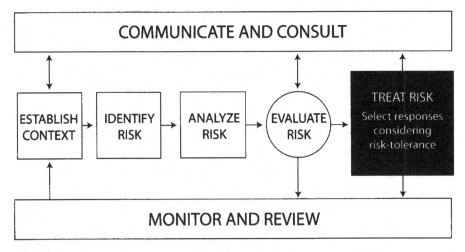

FIGURE 8.6
Make decisions to treat risks by including an assessment of risk tolerance.

Risk Acceptance

Accepting risk generally results when decision-makers deem the estimated level of risk to be tolerable and that the costs of mitigating the risk exceed the benefits gained. Discontinuing an activity may be a drastic response that is not feasible in the transportation sector. For example, shutting down an airport may minimize the risk of a terrorist attack to that facility but would have negative trade, commerce, and economic impacts (and may even transfer the risk to a different facility).

Risk Mitigation

Risk-mitigation measures can span a variety of organizational and tactical policies and programs. Countermeasures could serve to prevent an incident from occurring, detect an incident as it is in planning or action stages, delay the ability to perpetrate a hostile act, respond to an incident, and/or recover from an incident. Examples of possible risk-mitigation measures include:

- Prevent—in some instances it may be possible to institute measures to prevent a security incident; for example, enhanced information sharing between government security agencies could serve to identify and prevent a potential threat before it occurs.

- Protect—a number of measures are designed to protect people or structures against an incident; for example, instituting physical barriers or enhancing training of security personnel can reduce the vulnerability of facilities to some incidents.

- Respond—emergency operations plans for specific scenarios or incidents are examples of risk-mitigation measures for responding to an incident; establishing and practicing business-continuity or emergency-response planning can serve to minimize consequences from incidents.

- Recover—measures designed to improve the resiliency of a target are examples of risk-mitigation measures that facilitate recovery from an incident; for example, pre-incident planning can identify special-needs populations and help prioritize recovery efforts.

Ultimately, it is incumbent on identified decision-makers to evaluate the respective costs and benefits of each risk-mitigation activity in order to determine the most effective investment for their particular jurisdiction.

Making the Business Case for Risk Treatment

In order for maritime security and resilience measures to be fully embraced by all stakeholders, whether in the private or public sector, there needs to be a compelling business case.

This is important for the private sector in order to justify security expenditures that may increase the costs of doing business. By demonstrating the advantages to business continuity and loss prevention, a well-prepared business case, based on a sophisticated and rigorously analyzed risk assessment, can show the potential for security investments to reduce a private sector entity's risk exposure through risk treatment. In essence, an effective business case can help establish an organization's security and resilience functions as no longer just being cost centers but contributing to the organization's profits by reducing potentially harmful risks.

It is important to note that while private-sector acceptance of security and resilience measures and standards is desirable in order to

ensure wide acceptance and implementation, there are some requirements that will be imposed due to the differing responsibilities for
security that are held by government. The aim, however, should be
to make a business case whenever possible.

Counterintuitively, the process of developing a business case,
based on a clear understanding of the security and resilience risks
faced by an organization or sector and the most effective treatments,
can benefit government agencies as well.

Government agencies worldwide are under increasing pressure
to conserve funds due to tight budgets and the general atmosphere
of austerity as a result of the global economic downturn that began
in 2008. Further, some governments have been criticized for impos-
ing potentially crippling or draconian security regulations without
a full consideration of the potential implications on the maritime
sector or the ability of the private sector or government agencies to
execute the regulations. Therefore, developing a business case is a
natural component of a truly risk-based approach to maritime security in which identified risk treatments are tested against the variables
of cost, disruption, and potential efficiencies.

An example of a business case for the International Ship and
Port Security (ISPS) Code can be found in the results of a study by
the United Nations Conference on Trade and Development, which
asked port operators how they perceived the overall effect of the ISPS
Code.[1] Sixty-four per cent responded that the implementation of ISPS
had had a positive effect since it offered a mechanism to standardize
security in all parts of the port, while 61 percent noted that competitiveness was unaffected and 37 percent believed that their competitiveness had increased. Other important measures included the
assessment by 39 percent of respondents that their level of efficiency
had increased, while 61 percent believed that efficiency of operations was unchanged. While this information was compiled after the
implementation of the ISPS Code, it is indicative of some of the key
issues that may affect a particular risk treatment that should be considered prior to implementation if possible.

The business-case concept can be applied to any potential risk
treatment but should not necessarily in all cases be the most heavily weighted consideration. While preparing a business case can help

drive cost-effective solutions, some security and resilience solutions are simply going to be costly and perhaps inefficient but necessary. However, including a business-case analysis can help temper those solutions by demonstrating that no other, better options exist.

What is a Business Case?

A business case is a process that is used to determine whether a proposed course of action is appropriate for the organization involved. As the name implies, it originated in the corporate environment where proposed activities needed to be assessed against the key activities and requirements of the organization. The key issues considered in a business case pertaining to security and resilience issues can include:

- What is the cost?
- Can the cost be defrayed? If so, how?
- Are there ongoing maintenance, staffing, or training costs?
- Will the proposed enhancement to security or resilience contribute to the efficiency, reliability, or continuity of key functions?
- How does the proposed enhancement treat identified risks?
- What are the measures to determine the effectiveness of the enhancements?
- Is there a way for the enhancements to reduce operating costs?
- Is it required by law, regulation, insurance, or some other compliance body?
- How does it affect branding and reputation? Are there any potential advantages? Disadvantages? Can they be measured?

Regardless of whether the enhanced security or resilience measures comprise technology, physical security improvements, training, or programmatic issues such as planning and management, a business case should be included.

Composition of the Business Case

While there is no globally accepted template that delineates what should be included in a business case, there are several key considerations that should be included in a business case being used to justify security or resilience enhancement. Again, these components apply to government agencies as well as private-sector entities:

- The results of the latest risk assessment and an up-to-date risk register
- The expected costs of the enhancements
- Measures of effectiveness for the proposed changes
- A SWOT[2] Analysis of the potential consequences of the proposed enhancement, this is especially important when assessing the impact on interdependent sectors or systems
- The expected benefits in increased efficiency, reliability, compliance, and/or reputation

The Business Case and Risk Treatment

A well-structured business case provides a key tool in helping to determine the specific ways in which risk treatments that are identified in the risk-assessment process are implemented. The business case provides analytical rigor and requires consideration of a number of direct and indirect impacts as well as costs and measurement criteria. The key components of a risk-management program are laid out in

FIGURE 8.7

The business case as an essential part of risk management.

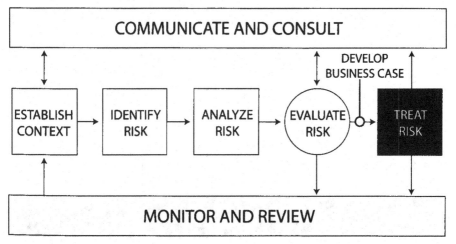

Appendix A and show the role of providing a business case in making risk-treatment decisions.

Monitor and Review

The purpose of this step in the risk-management process is to evaluate whether the strategies implemented to achieve security objectives and manage risk are and remain effective. This step is the key element of the PDCA approach in most international standards. The risk-management process seeks to measure risk levels as objectively as possible; however, any assessment will ultimately retain a level of subjectivity given that it is based upon human interpretation of information. Developing a transparent assessment process enables the analysis that risk ratings have been based on to be easily referenced, producing as accessible, rigorous, and reliable a risk-assessment process as possible.

For any risk assessment to be truly effective, it must be used as a "living document." A living document is one that is understood by all staff involved, achieved through training and regular updates. It should be tested on a regular basis through exercises, desktop scenarios, and real time workplace tests to fully appreciate the worth of the program. Each section of the risk assessment should be periodically and systematically reviewed and updated to ensure that it remains as relevant as possible and that the risk-management process can therefore be as effective as possible.

FIGURE 8.8

Monitor and review.

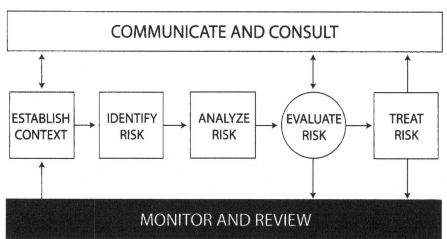

This approach ensures that risk management remains proactive. Stakeholders can utilize the principles of deliberate planning rather than being emergency- or response-driven. Potential activities in this stage include:

- Develop effectiveness criteria to report on performance and results in order to ensure risk-treatment measures are effective in design and operation.

- Analyze lessons learned from events, changes, and trends.

- Conduct regular and periodic strategic risk assessments to capture changes in the risk environment—emerging threats, diminished or increased vulnerabilities, etc.

- Record results to promote transparency and track performance over time.

- Review risk-analysis framework, consider methodological enhancements as data improves and other organizational capabilities grow and mature.

Communicate and Consult

Effectively communicating the results of the risk assessment and management process is essential to ensuring buy-in and cooperation with relevant stakeholders and partners in the implementation of risk-management measures. A mechanism and procedures should be put in place to capture the concerns and inputs of relevant internal and external stakeholders. Stakeholder engagement is important for several reasons.

Security planning, regulatory compliance, and operations are increasingly influenced by a complex environment of social, political, economical, and technological developments and interaction with stakeholders, each pursuing his or her own specific objectives. As a result, executives or decision-makers can experience stakeholder-management problems, which include:

- Communicating with a stakeholder too late in a regulatory process—this does not allow for ample revision of stakeholder expectations, and their views may not be taken into consideration.

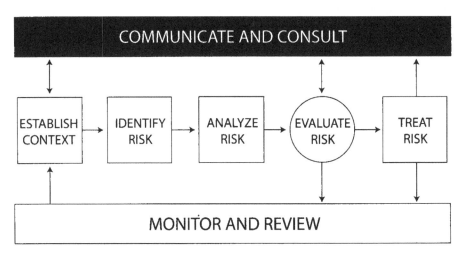

FIGURE 8.9

Communicate and consult to ensure stakeholder engagement.

- Inviting stakeholders to participate in decision-making process too early—this results in a complicated decision-making process.
- Involving the wrong stakeholders in a project—this results in a reduction in value of their contribution and leads to external criticism.
- Overlooking or underappreciating stakeholders—this can introduce significant variability to projects and result in project delay, added costs, lack of approvals, and reputational impact.
- Engaging stakeholders in isolation—this can result in stakeholder expectations and requirements that are contradictory or unnecessarily duplicative.

Therefore, it is vital that a stakeholder management plan involves a systems approach to security and an initially broad consideration of potential stakeholders. Once identified, a determination should be made regarding the relevance of the members of the stakeholder pool to the risk-management project under consideration. The pool should be as inclusive as possible, with compartmentalization kept to a minimum as required by the need to protect sensitive information. It is critical that key stakeholders are engaged in all elements of the risk assessment and management plan.

Maritime Considerations

The maritime operating environment is unique for numerous reasons documented in previous chapters. These include the fact that not all ports or maritime/intermodal transportation systems are equal. Each port area likely has a suite of issues that, taken together, form a unique context for that port. Therefore, risk assessments must be tailored to each port area, country, region, or sector. Further, where intermodal risks are included, the assessment process becomes far more complex, with a drastic increase in threats, stakeholders, and potential cascading effects. However, executed properly with appropriate stakeholder acceptance and input, a security risk assessment that provides risk-treatment options beyond simple mitigation can result in a focused application of resources on the key risks identified, which can in turn result in increased efficiencies in risk treatment and the reduction of costs for security and resilience efforts.

REFERENCES

1. United Nations Conference on Trade and Development, Maritime Security: ISPS Code Implementation, Costs and Related Financing, 14 March 2007, http://unctad.org/en/docs/sdtetlb20071_en.pdf, accessed, 2 June, 2012.
2. SWOT stands for Strengths, Weaknesses, Opportunities, and Threats. This is a common method of analyzing situations or potential courses of action.

CHAPTER 9

Measuring Effectiveness

INTRODUCTION

A major requirement for the implementation of effective maritime security and resilience measures is the need to be able to measure the effectiveness of implemented risk-treatment measures. A process to measure effectiveness is important in risk management in order to ensure that the initial measures are effective and to make adjustments as risks change or measures are shown not to be effective. Additionally, the ability to measure effectiveness is important in guiding investments in security and resilience to areas where the most significant impact can be realized.

The measurement of security, like the measurement of other disciplines where the prevention of incidents is the goal, is a complex endeavor and requires a combination of quantitative measures for threat scenarios involving large numbers of incidents and a qualitative approach to scenarios involving less common but potentially higher-impact incidents.

A major factor in assessing the effectiveness of preventive security measures is the level of deterrence that the measures provide against possible attackers or criminals. This consideration is particularly important as a key element of determining the effectiveness of security and resilience measures against low-frequency, high-impact incidents such as terrorist attacks and major crimes.

Measure Effectiveness, Not Security Activity

The initial shift required to accurately determine the effectiveness of preventive security and resilience measures is to focus on the effectiveness of the measures, not the activity involved in carrying them out.

As noted previously, the methodology to determine the effectiveness of maritime security measures by the Department of Homeland Security has focused on quantifiable metrics instead of qualitative, intelligence-derived assessments of the deterrent value. While not effective in preventing terrorist attacks, the quantitative approaches have strong foundations in both the bureaucratic and political environments and in conventional law-enforcement approaches to crime prevention and are valid in assessing the success of some maritime law-enforcement challenges including narcotics and migrant interdiction. However, these methodologies offer fundamentally flawed approaches to the measurement of the effectiveness of security measures against terrorist threats.

Currently, there is no established process to measure the effectiveness of preventive maritime security procedures outside of measuring activity or the possibility that there are classified reports regarding terrorist perceptions of security measures. This is part of the age-old quandary of trying to measure the ability to prevent something from happening, especially if it doesn't happen. Without the ability to measure effectiveness, it becomes challenging for security practitioners to assess priority targets for deterrence-based missions or the most appropriate security measures to be applied. Operational missions may include positive control boardings, vessel escorts, safety and security zone enforcement, security boardings, and customs inspections. Further, the Maritime Transportation Security Act (MTSA) and the International Ship and Port Security (ISPS) Code are designed to enhance security at waterside facilities and commercial vessels and are inherently deterrent-focused regulatory regimes. However, the degree of effectiveness of PWCS and MTSA measures has not been accurately determined. General information provided through the intelligence community has established the value of deterrence on the terrorist attack cycle, particularly during the surveillance and planning phases but there appears to have been no in-depth assessment regarding how effective certain tactics have been or which pieces of maritime infrastructure are best protected by which methods. In summary, detailed deterrence information is not available, which makes resource allocation problematic and complicates risk-based decision-making.

Measurement of Activity

The traditional measurement of activity in security operations takes two distinct approaches, a focus on resources expended and a quantitative approach to crime prevention.

Resources Expended

As noted previously, a common approach is to carefully document agency expenditures in order to justify current and future funding levels. This translates into a substitute for actually assessing the effectiveness of the activity but encourages lots of activity of indeterminate effectiveness. An excellent example of this is the legislation passed in 2007 requiring the screening of 100 percent of containers entering the United States. This legislation mandated a high level of activity but did not include any requirement to measure its effectiveness.

This measure is useful for those in government, as it is a simple way to document activity, especially if tied to plans that call for increased activity and resource expenditures as a means of deterring potential terrorist attacks. However, there is no analytic justification that ties increased activity to deterrence of terrorism.

Measurement of Criminal Activity

The second way that measurement is used in quantitative security-related analysis is the measurement of criminal activity. While this model may be useful for assessing the effectiveness of crime-prevention measures and, to an extent, border security, it is less effective in measuring the effectiveness of security against terrorist attacks on civilian infrastructure. There are two widely known tools for measuring crime in the United States.

Uniform Crime Reporting System

In the United States, crime figures are measured by the federal government by collecting data from law-enforcement agencies throughout the country and reporting the figures through the Uniform Crime Reporting (UCR) Program. While the UCR Program is very useful in determining detailed crime rates and trends, it does not provide any analysis of the reasons why crime rates may be rising or falling. Therefore, any analysis of the effectiveness of anticrime measures must be extrapolated from the data and is not directly correlated.

CompStat

Many municipal police departments use CompStat, an anticrime management approach that couples statistical analysis of criminal activity with the development and implementation of strategies to combat areas of high crime. It also incorporates a management approach that makes police managers accountable for the crime rates in their areas of responsibility and provides them with the ability to make decisions to combat crime rates. While generally perceived as effective, it has received criticism for pressuring police commanders to underreport crimes in their areas of responsibility, which, if true, harms the integrity of the data. There has also been speculation that crime-rate reductions that were attributed to CompStat in the 1990s may be the result of other factors, including an influx of new, better-educated police officers and the end of the crack epidemic, for example.

Unlike the UCR Program, CompStat is able to demonstrate a more direct, if imprecise, cause-and-effect relationship between crime rates and anticrime strategies. Therefore, under CompStat a degree of effectiveness can be measured.

The Black Swan Effect

The reason that a statistical approach to maritime homeland security is flawed, especially when measuring activity, is that statistical models that are geared towards budgetary concerns (activity reporting) or law enforcement (crime reporting) require significant amounts of data. Therefore, these models, while potentially effective in maritime homeland-security-related missions such as border security, narcotics, and illegal migration, do not make room for the "black swan" event that is a primary concern of maritime homeland-security practitioners. This theory was outlined in Chapter 4.

Measuring Effectiveness

The measurement of effectiveness of security measures can most effectively be carried out by the use of some of the established processes such as the UCR and CompStat methodologies coupled with a rigorous and more transparent assessment of the perceptions of security held by terrorist and criminal adversaries.

A Hybrid Solution

An effective method to measure the effectiveness of preventive security measures lies in the combination of a CompStat-type approach coupled with an intelligence-focused assessment of the adversary's perception of the effectiveness of security measures. In short, where possible, statistical information compared against security measures should be used, but we also need to evaluate the enemy through intelligence collection and analysis in order to be able to measure the effectiveness of security against potential high-impact, low-frequency events. This approach is well-suited for the ISO 31000-based risk-management system proposed in Chapter 8. This measurement of effectiveness, even if conducted on two levels, quantitative and qualitative, can provide essential information to an asset's security-management system and its cycle of continuous improvement.

FIGURE 9.1

Effectiveness measures and the PDCA cycle.

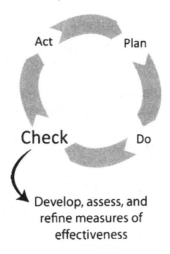

Ask the Enemy

Measuring the effectiveness of preventive maritime security activities focused against terrorism should center on the gathering and analysis of intelligence. Unlike measuring related law-enforcement-focused missions such as narcotics, smuggling, cargo theft, and other forms of criminal activity, measuring the effectiveness of counterterrorism measures cannot be assessed through statistical means, as there are insufficient seizures or arrests to provide a valid statistical assessment. This creates both a challenge and an opportunity. The challenge is that scientific models and statistical analysis are of limited utility, as adversaries will adjust their tactics and methods in response to security measures. Therefore, it is crucial that the assessment of effectiveness against potential terrorist acts is treated as an intelligence problem that analyzes the adversary's assessment of security and his or her intentions. In essence, "ask the enemy" what he or she thinks about the effectiveness of the maritime security program being analyzed.

Implementation

Clearly, an intelligence-based approach should be managed by governments that adopt this methodology. The collection segment of the intelligence cycle should be coordinated by the relevant intelligence agencies and should focus on the comprehensive collection plan developed in collaboration with nonintelligence security stakeholders. Further, well-constructed red-cell analysis by qualified participants, using the information provided, could be useful in identifying vulnerabilities and enhancing those security measures deemed to be most effective.

This collection effort should be designed to collect information from our adversaries using all intelligence disciplines. However, while there are several others, the key disciplines for this effort include:

- Human intelligence (HUMINT)—collecting information from human sources by human collectors through interviews of cooperating persons, recruited agents, confidential informants, prisoners, and friendly security personnel.
- Imagery intelligence (IMINT)—collecting information from sources of imagery, such as satellites, aircraft, and photography.

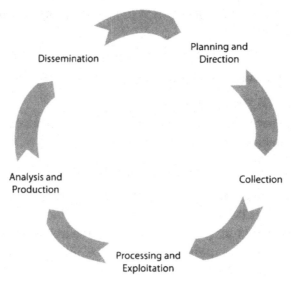

FIGURE 9.2

The intelligence cycle for maritime security. *Credit: US Central Intelligence Agency.*

- Signals intelligence (SIGINT)—collecting information from communications and electronic transmissions such as telephones, radios, and computers.

The Question

The development of intelligence requirements should focus on answering the primary question: are current maritime security measures effective against possible terrorist attacks? This fundamental question has a large number of possible subsidiary questions that can help develop more granularity in the collection plan and more fully determine the effectiveness of existing or prospective measures, including (but are not limited to):

- Have maritime or port assets been considered for attack?
- Why were they considered for attack?
- Were they rejected? If so, why?
- If they were not rejected, what type of preoperational surveillance and planning was conducted?

- How easy or hard was it to gather information on the target?
- How effective were the security measures in place?
- What was effective?
- What was ineffective?
- What influenced the decision to continue or not with attack planning?

While this area of maritime and port security is very sensitive and requires the protection of crucial information, it is important that key stakeholders be included in the process of identifying the collection requirements as well receiving the results, even in a sanitized format. Sources and methods need to be protected, but the results should be shared in a controlled atmosphere with key governmental and industry stakeholders. Accurately determining which preventive security and resilience activities are most effective in deterring our enemies will be useful at the national and local levels. Due to the ability of human adversaries to change their methods and tactics, the analysis of the effectiveness of our activities needs to be cyclical and recurring, which is consistent with the requirements of a security-management system.

Using Red Cells

A concern among some security professionals is that maritime security is suffering from "threat advocacy" in which elements of the international or national security community may have concerns that do not correspond with the adversary's actual strategic and tactical goals. This, coupled with the difficulty in conducting proper and effective threat analysis, has led to the acceptance of vulnerability-based security measures in which vulnerabilities are identified, a possible attack vector is identified, and a seemingly credible worst-case scenario is postulated. An example of "threat advocacy" explored earlier in this book focused on the requirement for the screening of 100 percent of all containers bound for the United States.

In the case of the threat of containers being used to smuggle radiological or nuclear material, the fact that it is physically possible for containers to be used to transport this material was sufficient for some political leaders to determine that the threat was viable rather than engaging in a threat assessment and using the results to carry out a risk assessment of the problem.

Therefore, in order to avoid "threat advocacy," red-cell activities should incorporate the mindset of the adversaries, which is a very specialized skill that requires the participation of individuals who have followed the particular adversary for years and are steeped in the culture and worldview. This is where the "if I were a terrorist" scenario development falls apart. Virtually every part of our society is vulnerable to terrorist activity, so using vulnerability-based analysis (even if it is cloaked in a "risk-based" approach) to develop proactive security measures is doomed to be incorrectly focused or overwhelming in its scope

There are flawed red-cell models in which former military special operators or even novelists and screenwriters are recruited. These professionals, the best in the world at what they do, can possibly provide information about the best way to attack a port, but it will be from a special-operators perspective or from a Hollywood producer's perspective, because they do not have the ability to look at the potential targets through the prism of a terrorist or a street-gang member. In the intelligence community, this is called "mirror-imaging," a situation where threats are assessed by individuals who are approaching the problem by looking at it through their own worldview, not that of their adversary. A retired special operator can provide the best information on how a commando would attack a port and identify vulnerabilities, but it is questionable whether that person could provide meaningful insight as to what the potential targets would be for an Al Qaeda operative and how Al Qaeda would carry out the attack. That is the work of intelligence analysts dedicated to following Al Qaeda.

Crunch the Numbers

Deterrence as the Primary Measure

In both quantitative and qualitative approaches to measuring effectiveness, the primary measure is the deterrent value security measures have or don't have. In pursuing an intelligence-based approach to measuring the effectiveness of security, the primary element to be evaluated is the deterrent value of the measures in place, and, if possible, proposed or potential measures. In the quantitative approach, deterrence will be self–evident by changes or reductions in criminal activity.

The qualitative approach is the most appropriate when assessing potential events that are infrequent but would have a very high impact, including most terrorist attacks. Conversely, a quantitative approach is suitable for events that are characterized as being of higher frequency but relatively lower impact. This is usually pertinent to assessing criminal activity.

Deterrence

As noted previously, deterrence is the art and science of preventing a person or organization from engaging in unwanted actions that threaten the asset or facility being protected and its ability to continue operations. There are three types of deterrence that are relevant to this problem:

- Punishment
- Denial
- Consequence management

Punishment

Punishment as a deterrent is based on the concept that the imposition of costs or punishment is sufficiently high to make an attack undesirable. The threat of punishment will not be particularly effective against suicide attackers, whose greatest fear is the fear of failure, not punishment, as they intend to die. Further, there is considerable debate among criminologists as to whether an enhanced threat of punishment serves as an effective deterrent against criminal activity.

Denial

Denial minimizes the opportunities available to adversaries. As a deterrent, denial focuses on taking measures to ensure the chances of adversary success are reduced or minimized. If the chances of an attack or criminal activity are reduced sufficiently, then the terrorists or criminals will be dissuaded from focusing on the facility, person, or organization being protected and will move on to targets that offer a greater chance of success. This form of deterrence is arguably the most likely to be effective against suicide attackers, as the fear of failure is the driving deterrent factor.

Consequence Management

Consequence management involves measures designed to minimize the impact of a successful breach or attack. This approach recognizes that the ability to convince an adversary that an attack or breach is not worth a risk are not successful but that the adversary will not gain a significant advantage by following through with an attack or criminal activity because the target will be able to minimize any impacts and will be able to recover quickly. Including consequence management as a deterrent provides a good basis to evaluate the resilience of a protected asset. A key indictor in measuring the potential effectiveness of consequence management efforts is the valuation of recovery times for key functions. This assessment, which is traditionally an element of business recovery, ensures that the key elements of an asset are protected and proper resilience planning is in place.

The three deterrent measures may be in many cases most effective, when combined and not implemented in isolation. The selection of these measures will most likely come as a result of the process used to develop a risk register as described in Chapter 8 and will be informed by the assessment of the level of risk that considered tolerable.

Ensuring Integrity and Countering Corruption

A key element in measuring the effectiveness of security measures is understanding the integrity of the system that is being protected from internal threats. These potential threats can include disgruntled employees, competitors, foreign intelligence agents, organized criminal groups, and terrorists seeking to sabotage or exploit parts of the maritime domain's infrastructure or supply chain. A key component of combating internal threats is the inclusion of a process to assess the transparency of all operations and activities. This includes interactions between governments and industry to ensure all activities are carried out appropriately and corruption is not permitted to fester.

Foster Continuous Improvement

As noted previously, the concepts of a systems-management approach to maritime security is key in ensuring that security and resilience measures are current and responsive to new threats or changes. The incorporation of a process to continuously measure the effectiveness of security and resilience measures is imperative to effectively determine the adjustments or modifications that may be required and any associated changes in resource requirements.

CHAPTER 10

Conclusion

This book is intended to provide maritime security practitioners with an overview of the current state of maritime and port security within the context of the maritime domain. It is further intended to provide a critique of current approaches to maritime security and recommended approaches to enhance maritime and port security. Those recommendations can be summarized as follows:

- Maritime and port security are not objectives but enablers to the lawful movement of goods and people.
- Maritime and port security are, by definition, an international concern. Unilateral approaches will rarely work.
- Technology provides tools, not solutions. If they are not deployed or used properly by humans, they are not effective.
- Threat assessments are crucial to a risk-based approach to security. If you don't know the enemy's intentions and capabilities, you can't adequately design security measures for best effect.
- Threats are assessed by people, not calculated. Purely quantitative approaches to assessing threats do not adequately consider the intent and capability of human adversaries.

- An all-hazard approach to security risk assessments and preventive measures seems efficient, but it leads to inaccurate risk prioritizations and ineffective risk registers because of the necessary lack of emphasis on threat in order to include nonsecurity hazards.

- An all-hazards approach to *response* can work effectively.

- A truly risk-based approach needs to include more than just risk mitigation. It should include all facets of risk treatment.

- In order to determine risk-appropriate risk treatments, the risk appetite or tolerance of the key stakeholders must be determined and validated.

- A truly risk-based approach to maritime and port security can lead to efficiencies that reduce the costs of security and increase competitiveness.

- The ISPS Code and global supply-chain security standards are compatible and should be combined.

APPENDICES

Conducting Security Risk Assessments

INTRODUCTION

This appendix provides guidance and templates for use by maritime security practitioners who are responsible for coordinating and conducting security risk assessments. The security risk assessment guidance and templates provided in this appendix are notional only and are consistent with the International Organisation for Standards 31000 on risk management.

This risk-assessment process provides practitioners with an understanding of the security risks present within their area of responsibility. Further, this process can be applied at both facility and strategic levels, allowing for in-depth understanding of security risk at individual maritime facilities or vessels, comparison of risk across facilities, and consideration of broader security challenges and potential risk-mitigation investment strategies within a more complex system such as a transportation sector.

Facility-level assessments involve site visits and operator interviews to collect facility-specific data that then can be aggregated and assessed. The results of facility-level assessments support the identification of security improvements at individual facilities as well as contribute to relative ranking of risk across facilities.

Strategic-level analysis can be conducted and involves analysis of generic representative targets (but not a specific facility). It can also integrate the analysis of trends across facilities. The benefits of strategic analysis include the ability to compare risk across transport modes or across critical infrastructure sectors in order to inform longer-term resource allocation and risk-mitigation investment strategies. Strategic risk analysis leads to the production of national aviation and maritime risk registers.

RISK ASSESSMENT STEPS

The risk assessment comprises elements of the ISO 31000, risk-management process, consisting of:

- Establish the risk-management context
- Identify risk scenarios
- Analyze risks, including consequences and likelihood of identified scenarios
- Evaluate risks by ranking risk scenarios relative to one another
- Treat risks by developing and implementing cost-effective strategies and action plans for increasing potential benefits and reducing potential costs
- Monitor and review risk and the effectiveness of treatment measures to ensure changing circumstances do not alter priorities

Of these elements, the first four comprise the risk-assessment section.

- Establish the risk management context
- Identify risk scenarios;
- Analyze risks, including determine consequences and likelihood of identified scenarios;
- Evaluate risks by ranking risk scenarios relative to one another;

The following overview provides a good introduction to the risk-assessment process.

Establish the Risk Management Context

This step of the risk-management process involves consideration of the goals and objectives, including the management and operational decision(s) to be supported by the risk assessment, the goals and objectives these decisions support, and an understanding of relevant decision-makers and other stakeholders, their security priorities, and their decision-making constraints. To ensure transparency of the risk-management process, it is important to clearly document the assumptions that supported the identification of the risk management context. As requirements change, it may be necessary to revisit these assumptions.

At the outset of a risk-management activity, it is useful to gauge the extent to which decision-makers are prepared to tolerate risk. Understanding decision-makers' risk appetite facilitates the development of strategies for prioritizing and mitigating risk.

Determining risk appetite can be accomplished through surveys, workshops, meetings with key stakeholders, research of documents, or a combination of the aforementioned. Some of the key elements to be considered in the determination of risk appetite include:

- Stakeholder knowledge of risk management
- Stakeholder comfort with risk
- Primary risks of concern
- Acceptable human death or injury losses
- Acceptable financial or economic losses
- Acceptable reputational damage
- Acceptable infrastructure damage
- Factors affecting risk appetite, including culture, security environment, operational environment, insurance, etc.

Identify Risks

Risk identification is the process of finding, recognizing, and describing potential risks. The identification of risks typically begins with consideration of what could go wrong, including noting targets or assets to be protected and consequences (loss of life, economic impacts, and reputational impacts) of concern to decision-makers.

Scenario development is a key step in the risk-identification process. Scenarios should be worst case-most likely case and contain a realistic depiction of potential threats. For security risk assessments, the identification and analysis of threats is covered in considerable detail in the following appendix.

However, in this section the key targets or assets to be protected and the consequences (loss of life, economic impacts, and reputational impacts) of concern to decision-makers need to be identified. This is done by identifying the critical operations of the asset or system to be protected.

Analyze Risks

The risk-analysis process should be designed to be flexible and repeatable in order to answer the primary three questions of security risk:

- What can happen?
- How likely is it to happen?
- What are the consequences if it does happen?

The sample risk-analysis methodologies in later sections of this appendix address these questions, analyzing the vulnerability and the consequence of each identified scenario as outlined by the ISO 31000 risk-management process. Threat and vulnerability together can be considered to determine likelihood, while consequence captures effects of interest to assessors and management.

It is useful to reiterate that hazards that come from accidents or natural disasters cannot be analyzed or assessed in the same manner as security threats. Therefore, this methodology focuses entirely on security threats.

Evaluate Risks

The purpose of risk evaluation is to assist in making decisions, based on the outcomes of risk analysis, about which risks need treatment in order to prioritize implementation. The evaluation of risk scenarios involves the ranking of these scenarios relative to one another to identify those that are of highest risk. It is often useful to develop risk registers to present the results of the assessment to decision-makers in an easily understood format. The results of risk assessments can be presented for individual facilities or aggregated across facilities to support the identification of trends and systemic risks to support strategic policy-making. Regardless of the presentation method selected, the results of the risk assessment and the underlying methodology should be fully documented to ensure the transparency and repeatability of the process.

CONDUCTING RISK ASSESSMENTS

Conducting a risk assessment of a facility, ship, or other entity involves careful preparation and proper expertise.

Assessment Team Composition

All Assessors

It is important that individuals involved in carrying out security risk analyses of maritime facilities or vessels possess relevant modal experience and formal assessor qualifications, including:

- Formal training in ISO 31000 and AS/NZS 4360 risk-analysis methodologies and approaches.
- Familiarity with international and relevant national maritime security standards, specifically the International Ship and Port Security (ISPS) Code.
- Familiarity with international supply-chain standards, specifically, ISO 28000 and/or WCO SAFE Framework.
- Knowledge of the maritime threat as it applies to the national security environment and the maritime sector.
- Formal internal or lead assessor training.
- A sound working knowledge of security technologies, techniques, and associated human factors within the maritime security environment.

Lead Assessor

The lead assessor is the appointed leader and manager of the assessment delivery who prepares the assessment plan, conducts the meetings, and write and submits the formal report.

The lead assessor is also the key liaison between the facility and the assessment team. Internally within the assessment team, the lead assessor is ultimately responsible for ensuring all administration and logistics are addressed before and during the assessment activity and that the assessors achieve individual and team objectives.

To achieve these objectives, it is essential that the lead assessor is a person who is fully capable of managing the delivery and coordination of the entire assessment process including the division of tasks to team members, managing individual team-member performance, and providing technical expertise within the team when required. In delivering the technical expertise, it is the responsibility of the lead assessor to provide a quality assurance of findings prior to completion of the analysis.

Assessment Team Members

Assessment team members are equally vital in ensuring team cohesion and achievement of individual and team analysis objectives. Analysis team members provide a level of assistance to the lead assessor by conducting meetings, noting observations, gathering objective evidence, and providing assessment findings to the team leader in accordance with methodology.

Facility Risk Assessment Process

A facility risk-assessment process involves four stages: preparation, site visit, reporting of assessment findings, and recording of assessment results. The site visit incorporates four key steps: review of documentation; discussion and interviews with assessed party at all levels; research into threat, vulnerability and consequences for the specific facility; and, observations of security delivery within seaport operations.

A consistent approach to risk assessment delivers quality and integrity and ensures that each assessment remains constant regardless of the assessor or assessed party. Adoption of this process allows for accurate assessment and review of results, thus facilitating the determination of risk and the effectiveness of risk treatments at an operational and strategic level.

Facility Risk Assessment Preparation

The assessment preparation process is a key function of the lead assessor and requires the coordinated efforts of both internal team members and the facility operators. The assessment preparation involves coordinating with facility operators, administration, and logistics; building the assessment team; and gathering of information for development of an assessment plan.

Written Notification to Facility Operators

As part of the preparation stage, written notification should be sent to facility operators months in advance, informing them of the scheduled assessment activity. The written notification should include the following:

- Purpose and scope of assessment
- Tentative date, time, and duration
- Details of assessment party, nominating lead assessor as point of contact
- Request for information on operations and change to operations
- Schedules
- Request for name and contact details of company personnel with security responsibilities
- Request for an onsite representative to be nominated to assist with scheduling of meetings
- Request for copies of security programs, manuals, procedures, and maps
- Details of any significant building and infrastructure work that has occurred since previous site visit
- Request for a private workspace and power
- Request advance approval for unfettered seaport access and access to relevant documents; this should be requested as early as possible in order to ensure that any security requirements are met
- Request advance approval to take photographs of findings and observations

The following is a comprehensive list of documents that will be of use to assessment teams:

- Site plans
- General layout
- Property borders
- Critical infrastructure, utilities, services, etc.
- Site boundaries and borders
- Routes into and out of the site
- Existing physical security controls (e.g., locations of CCTV units, checkpoints, fences)
- Location of sensitive operations and personnel
- Public and restricted access areas
- Emergency equipment (firefighting, first aid, breathing apparatus)
- Street maps
- External approaches
- Collateral exposures (e.g., neighbors, crowd areas, proximity to other higher-risk activities or infrastructure)
- Visibility and accessibility of the site
- Surveillance and overwatch vulnerabilities
- Policy and procedure documents
- Legislation
- Strategic and business plans
- Internal audit reports
- Business-continuity plans and test and exercise reviews
- Risk-management reviews
- Management and board reports
- Security-breach reports
- Broad incident trend data and intelligence from police and national security agencies

Planning Assessment Activity

Upon receipt of the requested information, and with confirmation of participating team members, the lead assessor is able to plan the assessment activity against functional seaport areas.

During the assessment period there will be a number of important activities, including the entry meeting and the closing meeting. When planning the site visit, it is preferable to agree to the day and time of the opening meeting, as this is usually the first activity.

When visiting a 24-hour site or operation, it is appropriate to split a team into two or more groups of two assessors. This approach ensures that all operations are reviewed and findings corroborated.

The safety and security access requirements should always be factored into the assessment plan. At some sites, a specific site safety and security induction is required prior to gaining any level of entry.

Of particular note, prior to the site visit date it is acceptable practice and professional courtesy to provide facility operators with a copy of the assessment program indicating task, team members, and assessment timings.

Facility Risk Assessment Administration and Logistics

In addition to planning the site visit, the lead assessor and team members must consider the administration and logistics that are required to successfully conduct an assessment. In this context, the following are to be considered:

- Travel/car hire
- Hotel
- Methods of payment for goods and services
- Occupational health and safety equipment
- Assessment tools
- Risk-assessment worksheets
- Camera
- Computer with printer if required
- Phone/satellite telephone if required
- Internet connectivity
- Passport photographs

Facility Risk Assessment Activity

The assessment contains four key activities that support the gathering of risk information: interviews/meetings, observations, document review, and research into facility-specific threat, consequence and vulnerability.

Document Reviews

The first activity that occurs within an assessment is generally the review of maritime security programs/plans and procedures. This activity can be conducted at the office before attendance onsite as part of the preparation/planning phase. During this activity, it is useful to detail the documents referenced on the checklist with appropriate commentary regarding conformity of the documentation to the benchmark standard. This review process is also used during the assessment activity to compare site manuals, programs, and databases.

Formal and Informal Interviews

Formal and informal interviews and meetings provide an avenue to determine the knowledge and competency of an individual as well as the level of implementation of security measures and perceptions of risk.

A formal or informal interview or meeting is best held in the presence of two assessors. An interview or meeting may be conducted anywhere, although a quiet location is more conducive for conversation.

Observations

Assessor observation of the implementation and delivery of protective security measures, such as viewing, conduct of security screening, parking controls, access control, wearing of security identification cards, port security signage, and baggage check-in procedures, provides an assessor with all or part of the evidence necessary to determine whether or not security-measure witnesses are compliant with the benchmark standard.

Assessment Opening and Closing Meetings

Opening Meeting

The assessment officially commences with an opening meeting attended by the entire assessment team, the facility security manager(s), the facility's senior management representative, and representatives of site security, police, customs, and immigration as appropriate. At the opening meeting the lead assessor serves as the only voice of the assessment team.

Closing Meeting

The closing meeting is the official end to the assessment activity. The same representatives generally attend.

Facility Assessment Reporting

At the completion of a facility assessment, assessors should submit the completed risk- assessment worksheets and any associated documentation to the team leader, who will in turn submit them to the IAC for collation, quality control, and initial assessment. The IAC will then make the information and initial assessment available to the risk department to support higher-level assessment.

Assessing Vulnerability

Vulnerability is a weakness in a target that can be exploited by the adversary. Vulnerability is assessed by evaluating the profile of the asset, taking into account its visibility and iconic nature, and the exposure of the target in terms of its accessibility and security countermeasures.

The profile reflects the likelihood that the adversary will be able to identify and locate the target, taking into consideration labeling or signage, press, uniqueness, and public knowledge of the asset. For attacks that require specific technical knowledge, the profile also takes into account the likelihood that an adversary can recognize or locate a specific critical location if necessary.

Exposure reflects the ease of access to the target as well as other security countermeasures in place to prevent the execution of a successful attack, including denial, detection, and interdiction measures. It also includes consideration of the target's ability to withstand the attack.

The individual vulnerability components can be combined to form a single measure of vulnerability. Profile and exposure can be combined to create a single index for the general vulnerability of a scenario. An example is shown in Figure A.3.

	Ranking Level				
Component	**Low**	**Medium-Low**	**Medium**	**Medium-High**	**High**
Profile	Asset is very unlikely to be recognized by adversary; an adversary would require a highly trained expert or access to classified or highly sensitive information.	Asset is unlikely to be recognized; an adversary would require some special knowledge or training.	Asset is somewhat likely to be recognized; an adversary would require a moderate amount of research.	Asset is likely to be recognized; an adversary could identify this asset with minimal effort.	Asset is very likely to be recognized; any adversary could easily identify this asset; attack method requires little to no direct targeting.
Exposure	Access to the asset is restricted and the existing counter-measures are very likely to defeat or withstand the attack.	Access to the asset is restricted and existing counter-measures are likely to defeat or withstand the attack.	Access to the asset can be achieved with minimal effort and the existing countermeasures are somewhat likely to defeat or withstand the attack.	The asset is publicly accessible and the existing counter-measures are unlikely to defeat or withstand the attack.	The asset is publicly accessible and existing security counter-measures are very unlikely to defeat or withstand the attack.

FIGURE A.1

Sample ranking table for vulnerability components.

PROFILE	EXPOSURE				
	Low	**Medium-Low**	**Medium**	**Medium-High**	**High**
High	Medium-Low	Medium	Medium-High	High	High
Medium-High	Medium-Low	Medium-Low	Medium	Medium-High	High
Medium	Low	Medium-Low	Medium	Medium	Medium-High
Medium-Low	Low	Medium-Low	Medium-Low	Medium-Low	Medium
Low	Low	Low	Low	Medium-Low	Medium-Low

FIGURE A.2

Sample vulnerability matrix.

Description	Consequence Components		
	National Reputation	**Economic**	**Human Health**
Low	No reputational damage with international partners with no resulting impact on flow of goods or people	Estimated loss from the incident are likely less than 1 million USD	Incident likely to cause no fatalities or injuries
Medium-Low	Minor damage to reputation with international partners resulting in limited degradation of flow of goods or people	Estimated loss from the incident are relatively minor, in the range of 1 – 10 million USD	Incident likely to cause no fatalities and some injuries
Medium	Moderate damage to reputation with international partners resulting in moderate degradation of flow of goods or people	Estimated loss from the incident in the range of 10-100 million USD	Incident likely to cause less than 10 fatalities
Medium-High	Significant damage to reputation with international partners resulting in significant degradation of flow of goods or people	Estimated loss from the incident in the range of 100 million to 1 billion USD	Incident likely to cause less than 100 fatalities
High	Major damage to reputation with international partners, resulting in long-term stoppages of flow of goods or people and degradation of international ratings of transportation facilities	Estimated loss from the incident in excess of 1 billion USD	Incident likely to cause greater than 100 fatalities

FIGURE A.3

Sample ranking table for consequence components.

Assessing Consequence

Consequence is the outcome of an event or change in circumstances affecting the achievement of objectives. Figure A.4 provides sample guidance for ranking consequence components. This ranking guidance is applicable for strategic or facility-level assessments.

The final consequence level reflects the cumulative effect of an attack or criminal act on fatalities, economic costs, and national reputational impact. Figure A.5 illustrates the consequence levels that derive from the combinations of the components.

Developing a Risk Rating

Once all of the information required has been collected and threats assessed, it is then combined to rate the security risks. Risk can then be expressed in terms of the likelihood that an attack will occur and the significance of the consequences to the asset if it does. This is then displayed as a risk register.

Sample Key for Aggregating Consequence Components

Component 1	Component 2	Component 3	Total Consequence Ranking
Low	Low	Low	Low
Low	Low	Medium-Low	Low
Low	Low	Medium	Medium-Low
Low	Low	Medium-High	Medium-Low
Low	Low	High	Medium-Low
Low	Medium-Low	Medium-Low	Medium-Low
Low	Medium-Low	Medium	Medium-Low
Low	Medium-Low	Medium-High	Medium-Low
Low	Medium-Low	High	Medium
Low	Medium	Medium	Medium-Low
Low	Medium	Medium-High	Medium-High
Low	Medium	High	Medium
Low	Medium-High	Medium-High	Medium
Low	Medium-High	High	Medium
Low	High	High	Medium-High
Medium-Low	Medium-Low	Medium-Low	Medium-Low
Medium-Low	Medium-Low	Medium	Medium-Low
Medium-Low	Medium-Low	Medium-High	Medium
Medium-Low	Medium-Low	High	Medium
Medium-Low	Medium	Medium	Medium
Medium-Low	Medium	Medium-High	Medium
Medium-Low	Medium	High	Medium
Medium-Low	Medium-High	Medium-High	Medium
Medium-Low	Medium-High	High	Medium-High
Medium-Low	High	High	Medium-High
Medium	Medium	Medium	Medium
Medium	Medium	Medium-High	Medium
Medium	Medium	High	Medium-High
Medium	Medium-High	Medium-High	Medium-High
Medium	Medium-High	High	Medium-High
Medium	High	High	Medium-High
Medium-High	Medium-High	Medium-High	Medium-High
Medium-High	Medium-High	High	Medium-High

FIGURE A.4

Example of rating consequence components.

The three factors of risk, threat, and vulnerability can be combined to provide an estimate of the likelihood of an event occurring. This is then combined with consequence. The combination of the resulting likelihood level with the consequence level allows the generation of a risk level. This matrix can then be developed and organized into a prioritized risk register. An example of a risk register can be seen on the next page in Figure A.7.

THREAT	VULNERABILITY				
	Low	Medium-Low	Medium	Medium-High	High
High	Probably not	Chances about even	Probable	Almost Certain	Almost Certain
Medium-High	Probably not	Probably not	Chances about even	Probable	Almost Certain
Medium	Almost certainly not	Probably not	Chances about even	Chances about even	Probable
Medium-Low	Almost certainly not	Probably not	Probably not	Probably not	Chances about even
Low	Almost certainly not	Almost certainly not	Almost certainly not	Probably not	Probably not

FIGURE A.5

Sample likelihood matrix.

LIKELIHOOD	CONSEQUENCE				
	Low	Medium-Low	Medium	Medium-High	High
Almost Certain	Medium-Low	Medium	Medium-High	High	High
Probable	Medium-Low	Medium-Low	Medium	Medium-High	High
Chances about even	Low	Medium-Low	Medium	Medium	Medium-High
Probably not	Low	Medium-Low	Medium-Low	Medium-Low	Medium
Almost certainly not	Low	Low	Low	Medium-Low	Medium-Low

FIGURE A.6

Risk matrix.

Facility	Scenario	Threat Ranking	Vulnerability Ranking	Consequence Ranking	Likelihood Ranking	Risk Score

FIGURE A.7

Example of a risk register.

Conducting Threat Assessments

INTRODUCTION

The assessment of security threats provides a key component in identifying risks. This tool is designed to be consistent with the risk-assessment methodology developed for ISO 31000, which provides a standard for risk management. Threat assessment is critical in developing a security risk assessment by identifying the source of the risk. This is usually accomplished by assessing two major factors affecting threat; intent and capability. However, in order to develop a rigorous and defensible threat assessment, the following factors need to be included: existence, history, targeting, and the security environment.

The process of assessing threats requires a significant amount of analytical rigor and a clear process to ensure that threats and their associated scenarios are correctly identified. Further, threat assessments are perishable due to the regularly changing nature of threat groups and their associated targets and tactics and should be reevaluated on a periodic basis unless a significant security event occurs or any of the key variables in the threat assessment change, spurring a more frequent reevaluation. It may be useful for the practitioner to develop a set of potential indicators or events that would trigger a more frequent threat-assessment review.

CONSISTENCY WITH ISO 31000

This methodology is designed to be consistent with the steps and elements of ISO 31000, with its focus on continuous improvement. ISO 31000 is a process that comprises the following steps and elements:

- Establish the context in which the risk is being assessed
- Identify the risks which the organization may face
- Analyze the risks
- Evaluate the risks
- Treat the risks

Throughout the risk-assessment process, there are two further critical elements of the risk-management process:

- Communication and consultation with all relevant internal stakeholders to ensure the support of senior management and participation of relevant personnel as well as the determination of risk tolerances.
- Monitoring and reviewing the process to ensure that assessments and recommendations remain relevant in light of any changes in the context or risks.

FIGURE B.1

ISO 31000.

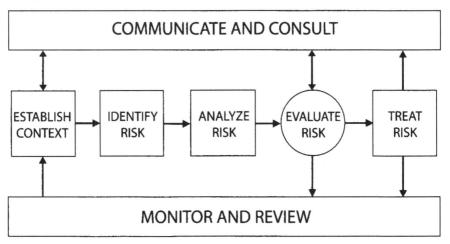

The components of this threat-assessment methodology include the following:

- Threat identification
- Identify the range of potential threat actors
- Identify an extensive list of threat actor characteristics
- Identify sources of threat-related information
- Analyze and organize threat-related information
- Threat evaluation
- Threat actors and scenarios
- Develop the design basis threat

Figure B.2 demonstrates the consistency between the ISO 31000 methodology and this threat-assessment methodology, including the built-in focus on continuous improvement and the need to regularly review the threat assessment.

Threat Identification

This stage is similar to the requirements in a risk assessment to determine the context, or factors in establishing which sorts of threats should be included in the assessment. Threat identification focuses on the actors that are responsible for the threat. As threats, unlike hazards, are caused by cognizant human adversaries, they will always have an actor. Threat actors may be individuals or groups.

FIGURE B.2

Threat-assessment methodology compared to ISO 31000.

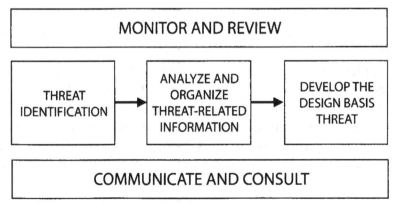

Identify the Range of Potential Threat Actors

Recommended categories of generic adversary threats include:

- External threat actors
- Protestors, including demonstrators, activists, and extremists
- Terrorists
- Criminals, including organized crime
- Internal threat actors
- Insiders: disgruntled employees, labor agitators, corporate spies
- Potential industrial saboteurs
- Employees or corporate leadership interested in committing fraud or violating laws for personal or corporate gain

The processes outlined in the remainder of this section are generally applicable, regardless of the threat category being considered.

Identify an Extensive List of Threat Actor Characteristics

For each of the threat actors identified, information must be gathered to define the threat. This step represents the information requirements associated with the development of the threat assessment. At a minimum, information must be gathered regarding motivation/ intention and the following list of capabilities: number of attackers, weapons/explosives, tools and equipment, transportation, technical skills and expertise, funding, collusion with insiders, and knowledge of the facility and its operations.

Identify Sources of Threat-Related Information

Once collection requirements for threat actor characteristics have been established, sources that can provide the required information must be identified and exploited. A partial list of open sources that can be consulted regarding the terrorist threat has been provided below. Additionally, threat analysts may have sources within governments and security agencies that can provide relevant information.

Further, commercial concerns that may be either targets or victims of potential threat actors may provide valuable information.

An issue that arises frequently is the concern that a threat assessment cannot be performed without access to classified information. While in some cases this is true, in most cases there is sufficient open-source information to allow an experienced and dedicated analyst to identify the intent and capability of threat actors.

The list below is not exhaustive and does not reflect any regionally specific information sources that may be of utility to a threat analyst. Further, the list covers a wide variety of news sources, each of which has some bias that should be carefully considered when conducting threat analysis:

News sources/media

- CNN: www.cnn.com
- British Broadcasting Company: www.bbc.co.uk
- Reuters: www.reuters.com
- Associated Press: www.ap.org
- *New York Times:* www.nytimes.com
- *Washington Post:* www.washingtonpost.com
- Press TV: http://www.presstv.ir/
- *The Guardian:* http://www.guardian.co.uk/
- *Al Arabiya:* http://www.alarabiya.net/english/
- *Al Jazeera:* http://www.aljazeera.com/
- *The Jerusalem Post:* http://www.jpost.com/
- *Russia Today:* http://rt.com/
- *Xinhua:* http://www.xinhuanet.com/english/
- News Link: www.newslink.org—repository of worldwide news media sources, including newspapers, TV, radio, etc.
- Ref Desk: http://www.refdesk.com/paper.html—repository of US and worldwide newspapers

Research centers/academic institutions

- The Centre for the Study of Terrorism and Political Violence, St. Andrew's University: www.st-andrews. ac.uk/~wwwir/research/cstpv/

- Combating Terrorism Center, United States Military Academy: www.ctc.usma.edu/
- The RAND Corporation: http://www.rand.org/research_areas/terrorism/
- The Middle East Media Research Institute: www.memri.org —focuses on covering developments in the Middle east and Central Asia
- Maritime Terrorism Research Center: http://www.maritimeterrorism.com/—provides information and intelligence regarding maritime terrorism and maritime security
- Terrorism Research Center, Fulbright College, University of Arkansas: http://trc.uark.edu—conducts research to better understand the spatial and temporal patterns of terrorist activities
- Center for International Research on Terrorism: www.terrorismresearchcenter.org—independent nonprofit organization focused on scholarly research and public education on terrorism affairs based on thirty-year counterterrorism experience of Turkish National Police
- Terrorism Research Center: http://www.terrorism.com/index.php—provides research and information on terrorism, information warfare and security, critical infrastructure protection, homeland security, and other issues of low-intensity political violence
- Center for Defense Information: Terrorism Project: http://www.cdi.org/terrorism/terrorist-groups.cfm—provides information on international terrorist organizations considered by the US Department of State to be active within the past five years
- Center for Intelligence Research and Analysis: http://www.defensegroupinc.com/cira/index.cfm—provides open-source and cultural intelligence services to clients worldwide
- Southern Poverty Law Center: www.splcenter.org— provides information on hate groups and some terrorist groups, primarily in the United States

US Government sources

- US Library of Congress Federal Research Division Country Studies: http://lcweb2.loc.gov/frd/cs/
- US Library of Congress Portals to the World: http://www.loc.gov/rr/international/portals.html
- US National Intelligence Council: http://www.dni.gov/nic/NIC_home.html
- US National Counterterrorism Worldwide Incidents Tracking Center: http://wits.nctc.gov/

Commercial/industry sources

- Security Debrief: www.securitydebrief.adfero.com—blog on homeland-security affairs providing descriptions of terrorist attacks worldwide
- Memorial Institute for the Prevention of Terrorism: www.mipt.org/—counterterrorism center focused on training, analysis, and information sharing on the prevention of terrorist attacks
- The Intel Center: http://intelcenter.com/—provides updated information about worldwide terrorist attacks and warnings
- JIOX Intelligence and Tradecraft Analysis: http://jiox.blogspot.com—contains a bounty of strategic intelligence reporting and a thorough list of open-source links
- ERRI Counterterrorism Archive: http://www.emergency.com/cntrterr.htm—provides a summary of worldwide terrorism events, groups, strategies, and tactics
- Stratfor: www.stratfor.org—provides strategic intelligence on global business, economic, security and geopolitical affairs.
- Jane's Information Group: www.janes.com—provides news, information, and analysis from leading sources on defense, geopolitics, transport, and police industries and issues
- Intelligence Online: http://www.intelligenceonline.com/—provides reports and assessments on political intelligence issues

Analyze and Organize Threat-Related Information

This step requires the analysis of open-source information gathered during the collection process in order to define the range and characteristics of the terrorist threat. Its end state is the creation of a draft threat assessment with its roots in a design basis threat (DBT), which is more fully explained in a later section of this appendix. In an effort to guard against an overstatement of the threat, the DBT should be developed through rigorous research that results in scenarios and threat actors that meet the requirement that "the threat must be credible." Here, "credible" means demonstrable, and threat capabilities that have not been demonstrated or shown to be possible in real-life events should be excluded from the threat assessment. While it is important to consider potential future scenarios and capabilities, the DBT must be realistic in order to make informed decisions about the allocation of finite resources for physical protection systems. A DBT sample format is found in Figures B.7–B.14.

Threat Evaluation

Threat groups are evaluated against the following criteria:

- Does this organization currently have a presence within the city, country, or region?
- Is this organization associated with other organizations, whether legal or not?
- Has this group engaged in attacks/criminal activities in the city, country, or region?
- Have the attacks/criminal acts had a maritime, transportation or energy nexus?
- What did the activity consist of?
- Is the group capable of carrying out attacks or criminal activities?
- Is there logistical/popular support for the group?
- Are there ties to other capable state or nonstate actors that can influence capability?
- How robust are current security precautions to guard against attack or criminal activity?

Negligible	No history, skills, resources or knowledge and as a result, no capability to realize threat.
Minimal	Has minimal access to the resources, skills and knowledge required to realize the threat. Small and probably ineffectual support base.
Limited	Has minimal access to the resources, skills and knowledge required to realize the threat, or part of the threat. Limited support base.
Partial	Has access to the resources, skills and knowledge to realize the threat, if not in direct possession of such resources.
Significant	Probably has the requisite resources, skills and knowledge to realize the threat under certail circumstances. Has a committed although not necessarily extensive cadre of support.
Complete	Has not recently demonstrated the presence of resources required to realize the threat, but is likely to possess the required resources. has access to a solid support base and skills required to realize the threat.
Substantial	Recent evidence of being well resourced with the requisite skills and knowledge to realize the threat. Evidence of training and organizational ability and has access to a solid and extensive support base.

FIGURE B.3

Threat capabilities assessment.

- What is the capability of the security forces to combat the threat?
- What is the political will of the security apparatus to combat the threat?
- Is there widespread corruption among the security forces?
- What is the articulated goal(s) of the organization?
- What is the actual goal(s) of the organization if different?

Negligible	No desire to cause harm. Consequences of causing the threat to be realized would be counter-productive and no indication that the subject has any expectation of causing the consequence.
None-evident	No evidence of a desire to cause harm and no history or record of having caused the consequence envisaged, but might attempt to do so in extreme circumstances.
Suspected	Could have a desire to cause harm and might have some expectation of successfully causing the consequence envisaged.
Reasonable	Would reasonably have the desire to cause harm and the consequences would be consistent with the overall aims of the subject.
Probable	Anecdotal or other evidence to suggest there is an intention to cause harm and actions would be consistent with overall aims.
Stated	Has made clear statements that indicate a desire to cause harm and some confidence of being able to carry out threat.
Demonstrated	Has articulated and demonstrated a public and determined desire to engage in harm against the subject in order to achieve goals. Has had success in the past that will result in expectation of future successes.

FIGURE B.4

Threat intent assessment.

- What is the intended outcome of the attack or criminal activity (e.g,. loss of life, economic damage, reputational damage, use of port areas as a conduit)?
- Is there evidence (either articulated or actual) of active targeting?
- Was the primary target infrastructure or personnel?

Using the information collected and organized above, the capabilities and intentions/motivations can be assessed using the criteria found in the following figures. These values can be combined to arrive at a threat rating. The threat rating can be derived from the table shown in Figure B.5.

Figure B.5 brings all elements of the threat assessment together in order to allow each asset being assessed as part of a facility or system to have threat ratings applied when various scenarios are considered.

Threat Actors and Scenarios

The following figures are sample threat scenarios and potential actors developed using a modified design basis threat (DBT) methodology for notional seaports. These figures provide the user with documented capabilities and scenarios that can be used to assess threats. This structured approach allows the user to focus on established and appropriate threat vectors rather than on an almost unlimited array of possible scenarios that have not been subjected to analytical rigor.

Develop The Design Basis Threat

A DBT is a profile of the threat that any facility, industry, or government should be prepared to defend against. As such, the DBT serves as the basis for the design and evaluation of a protection system. It characterizes the elements of a potential attack or criminal event, including the number of attackers, their training, and the weapons and tactics that they are capable of employing. It is important to note, however, that the DBT does not represent the maximum possible size and capability of an attack but rather the characteristics of an attack that pose an unacceptable risk to the facility or infrastructure at hand. As its development is based on an analysis of historical events, it usually represents the current threat rather than the forecasted threat. As such, the DBT should be seen as a living document that must be updated in a consistent and deliberate fashion to ensure the adequate protection of critical infrastructure. As stated above, it is recommended that the DBT be updated at predesignated intervals and more frequently if necessary.

The development and use of a design basis threat is important for three main reasons:

1. The DBT is the basis for the design of a protection system, whether it is a national policy or a facility security plan. If security executives do not know who the adversary may be or the capabilities and intent that the adversary may possess, then the design and implementation of the protection systems are not likely to be effective or focused appropriately.

2. The DBT is the basis for evaluation of a protection system or plan. The evaluating party can use the DBT to evaluate the effectiveness of a security program against the threat. Without a DBT there is no objective measure for evaluating a program's effectiveness. This could lead to inconsistent evaluations and gaps in protection.

3. The DBT provides a framework to document future threat changes. Threats against facilities, infrastructure, and nations change constantly, with some increasing and others decreasing. If the threat is not periodically and rigorously assessed, these changes could pass undetected. Developing a DBT provides the standard against which future threats are compared.

The following Figures B.7 through B.14 provide examples of generic potential actors and associated scenarios that give an overview of the value and level of detail possible in a DBT. These figures were developed using open-source information. They are examples that will need to be tailored to specific threats and should be considered illustrative only.

Capability	Intent						
	Negligible	None Evident	Suspected	Reasonable	Probable	Stated	Demonstrated
Substantial	Medium	Medium	High	High	Very High	Very high	Extreme
Complete	Low	Medium	Medium	High	High	Very High	Very high
Significant	Low	Low	Medium	Medium	High	High	Very High
Partial	Very Low	Low	Low	Medium	Medium	High	High
Limited	Very Low	Very Low	Low	Low	Medium	Medium	High
Minimal	Nil	Very Low	Very Low	Low	Low	Medium	Medium
Negligible	Nil	Nil	Very Low	Very Low	Low	Low	Medium

FIGURE B.5

Threat rating table.

Infrastructure Risk Assessment
*Sample **Terrorism** Threat Assessment Worksheet*

Date:	
Infrastructure:	
Representative Name and Title:	
Assessor Contact Information (Phone/E-mail):	
Threat:	

Terrorist Attack Methods

Airborne Attack	Cyber	Food or Water Contamination	Standoff Weapons
- Cargo	Covert Intrusion	Hijacking	- Direct Fire
- Passenger	Explosives	RDD	- Indirect Fire
Biological Disease	- IED	Sabotage	Storming
Chemical	- Maritime IED	Small Arms	
- Toxic Industrial	- Explosives-laden boat		
- Warfare Agent	- VBIED		

Description of worst-most likely scenario:
Scenario should be in collaboration with assessors of vulnerability and consequence.

Asset	

Significance or criticality of the infrastructure asset:
Description should be in collaboration with assessors of vulnerability and consequence.

FIGURE B.6

Sample threat-rating worksheet *(this figure continues on the next page).*

Threat Rankings (Terrorism)	Man-made occurrence, individual, entity, or action that has or indicates the potential to harm life, information, operations, the environment, and/or property.[i] The terrorist threat ranking is assessed by **capability** and **intent** of the adversary.
Threat Actor:	
Assessed Current Capability Ranking (*check ranking, utilizing ranking guidance*) ☐ Negligible ☐ Minimal ☐ Limited ☐ Partial ☐ Significant ☐ Complete ☐ Substantial	**Reason for ranking current capability of adversary to conduct described attack:** *(Please source reports, studies, or information supporting ranking.)*
Intent Ranking: *(check ranking, utilizing ranking guidance provided)* ☐ Negligible ☐ None Evident ☐ Suspected ☐ Reasonable	**Reason for ranking intent to described attack:** *(Please source reports, studies, or information supporting ranking.)*

FIGURE B.6 *(continued)*

Sample threat-rating worksheet.

Adversary	Source	Threat History
XXXXXXX	• Most likely to be either resident expatriates, visitors possibly supported by external intelligence officials.	• Focused on xxxxxxx and immediate environs. • Also believed responsible for some attacks in xxxxxxx. • Historical focus on U.S. and other Western persons, military, embassy assets in xxxxx, including the bombing of the xxxxxxxx. • Reportedly training xxxxxxxxx insurgents in xxxxx.
XXXXXXXXX	• Both domestic and foreign nationals either resident or visiting.	• First attacks described in xxxxxx. • Historical focus on aviation, mass transit, western tourism/business interests, military, and embassies • In xxxxxxxx, terrorists attacked several iconic targets in xxxxxx.

FIGURE B.7

Terrorism threat.

Potential Actions/Scenario*	Adversary Capability	Adversary Intent/Motivation
1. Small arms assault 2. Explosive attack (pedestrian or VBIED) 3. Stand off attack (i.e. RPG, surface to surface missile) 4. Airborne attack (Kamikaze) 5. Covert intrusion 6. Paramilitary small unit operations 7. Training of indigenous groups * Detailed scenarios are outlined in Tables xx and xx.	• Established capability in xxxxxxx. • Definite logistical support base and possible operational capability in Europe and N. America. • State sponsors include xxxxx. • Likely to have capability in the xxxxxxx. • Specific capability in xxxxxxx is unknown but if relations with xxxxxx deteriorate, likely to establish/increase a presence.	1. To expel U.S. and other "colonialists" from xxxxxxx. 2. To establish a xxxxxx government. 3. Carry out operations on behalf of xxxxxx. 4. Objectives may be modified due to xxxxxx increased involvement in civil and political life.
1. Small arms assault 2. Explosive attack (pedestrian or VBIED) 3. Stand off attack (ie. RPG) 4. Airborne attack (Kamikaze) 5. Covert intrusion * Detailed scenarios are outlined in Tables xx and xx .	• Established capability in the region. • Assessed to be regaining some strength for larger scale attacks. • Recent successes by xxxxx security forces have diminished but not eliminated its support base and operational capabilities.	• To expel U.S. and western military from the region. • Attacks against oil infrastructure would satisfy stated objectives to undermine western presence and stability of the government.

Adversary	Source	Threat History
Internal conspirators	Employees or contractors	• In xxxxxxx, crewmembers of a xxxxxxxx-flagged ship were arrested in xxxxxx and were charged with conspiracy to commit terrorist acts.
Industrial espionage	Employees or contractors	• Industrial espionage has a long history and is primarily focused on the theft of trade secrets and the development of a competitive advantage. • In xxxxx, the xxxxx arrested xx individuals and charged them with industrial espionage by trying to obtain trade secrets on behalf of foreign oil companies.

FIGURE B.8

Internal espionage/labor unrest threat.

Potential Actions/Scenario*	Adversary Capability	Adversary Intent/Motivation
1. Insider threat 2. Small arms assault 3. Covert intrusion 4. Explosive attack 5. Stand-off attack * Detailed scenarios are outlined in Tables xx and xx.	• Likely to be able to develop the capability rapidly if it doesn't currently exist. • Capability clearly extant in the region.	• Sabotage due to dissatisfaction. • As directed by outside sources.
1. Insider threat	• Many companies have competitive intelligence components of varying degrees of proficiency. • Some national governments also have industrial espionage programs designed to assist their national industries.	• To obtain a competitive advantage.

Adversary	Source	Threat History
Expatriate Labor Agitators	• Foreign nationals working at oil/gas facilities or related transportation sectors.	• Recent labor unrest is partially attributed to increased inflation.

FIGURE B.9

Internally assisted espionage/labor unrest threat.

Adversary	Source	Threat History
International Environmental Activist Groups	• Persons who are members of environmental activist or anti-globalization groups.	• No specific actions in the region but articulated concerns regarding the contributions of petroleum products to global warming. • Anti-globalization groups have not been engaged in any overt activities in the region.

FIGURE B.10

Environmental activism threat.

Potential Actions/Scenario*	Adversary Capability	Adversary Intent/Motivation
1. Covert intrusion 2. Labor Unrest 3. Insider Threat * Detailed scenarios are outlined in Tables xx and xx.	• Currently capable at relatively small scale (ie. Single sites). • No known broader organization. • Unrest has largely been in the construction sector.	• To improve wages and guarantee timely payment of wages. • To improve living conditions. • Possibly corporate espionage.

Potential Actions/Scenario*	Adversary Capability	Adversary Intent/Motivation
1. Covert intrusion 2. Insider Threat * Detailed scenarios are outlined in Tables xx and xx.	• Not currently believed to have any level of capability in the region.	• To bring attention to environmental issues ranging from global warming to animal rights, to the production of genetically modified foods. • To protest globalization policies that are perceived as harmful to the developing world.

Adversary	Source	Threat History
Organized Criminal Entities	• Members of organized criminal groups. • Ad hoc criminal groups.	• History of xxxxx ports being used to transship controlled technologies from other nations to xxxxxx. • History of xxxxxxx ports serving as a transshipment point for narcotics heading to Europe or the United States, • Presence of xxxxxxxx organized criminal groups exacerbate smuggling and trafficking of narcotics, people, and other contraband. • Maritime piracy and vessel hijacking has not occurred in xxxxxx waters in recent history. Apparent hijacking of the xxxxxx in the xxxxxx by what appear to be organized crime elements indicates a broader level of targeting and possible realization of significant monetary gain.

FIGURE B.11

Trafficking/smuggling/hijacking threat.

Adversary	Source	Threat History
Potentially Hostile State Actors	• xxxxxxxxx	• xxxxx has historically been at contretemps with the xxxxxxx. • xxxxxx has aggressively projected its naval forces in the xxxxxxx, encroaching on the territorial seas of neighboring countries and challenging foreign naval assets.

FIGURE B.12

State actors threat.

Potential Actions/Scenario*	Adversary Capability	Adversary Intent/Motivation
1. Covert intrusion 2. Vessel Hijacking * Detailed scenarios are outlined in Tables xx and xx3.	• Smuggling remains problematic as there is a relatively long coastline and land borders with minimal controls. • Traffickers have the capability to exploit regional maritime traffic as well as larger commercial vessels. • Merchant vessels are slow, vulnerable targets, especially in waters where vessel hijackings are not common and regional maritime security forces may not be able to immediately respond.	• Primarily profit-driven with possible ancillary political objectives.

Potential Actions/Scenario*	Adversary Capability	Adversary Intent/Motivation
1. Insider Threat 2. Small Arms Assault 3. Covert intrusion 4. Explosive Attack 5. Stand-Off Attack (surface to surface) 6. Airborne Attack 7. Conventional military assault, both land and maritime * Detailed scenarios are outlined in Tables xx and xx.	• xxxx has a substantial military presence in the region through its conventional military forces as well as the xxxxxxx. • xxxxxxxx capabilities include small boats and special operations forces.	• Geo-political with a focus on regional hegemony. • xxxxxxx: Potential territorial disputes.

	Small-Arms Assault	Covert Intrusion	Explosive Attack (Pedestrian/ Cargo/Vehicle-Borne)
Attackers	• 1-12 attackers • Knowledge of facility through open source information, observation, and/or insider collusion • Trained in the use of explosives and small arms	• 1-5 attackers • Knowledge of facility through open source information, observation, and/or insider collusion • Trained in the use of explosives and small arms	• 1-5 attackers • Knowledge of facility through open source information, observation, and/or insider collusion • Trained in the use of explosives and small arms
Transport	• Pedestrian • Minimum of 2 vehicles; • Off-road all-terrain vehicles • Motorcycles • Cars, vans, mid-size trucks • Vehicles may be painted/ marked to appear to be official (e.g. ambulance, police) or company	• Pedestrian • Off-road all-terrain vehicles • Motorcycles • Cars, vans, mid-size trucks • Vehicles may be painted/ marked to appear to be official (e.g. ambulance, police) or company	• Pedestrian • Minimum of 2 vehicles; • Off-road all-terrain vehicles • Motorcycles • Cars, vans, mid-size trucks, large cargo /container trucks • Vehicles may be painted/ marked to appear to be official (e.g. ambulance, police) or company
Weapons / Equipment	• Pistols, submachine guns, assault rifles, sniper rifles (up to .50 caliber), light machine guns • Rocket propelled grenades (RPG) • Grenades (high explosive and incendiary) • Bulk explosives for access or attack (up to 4500kg/10000lbs TNT equivalent depending on vehicle) • Specialized explosive charges (breaching charges, shape charges,, etc.) • Chemicals (chlorine, toxic gases) • Official or corporate uniforms and credentials	• Mechanical breaching tools, quick saws, chainsaws, sledge hammers, hand tools, ladders/ropes/material to cover topguard • Grenades (high explosive and incendiary) • Bulk explosives - 30kg/65lbs TNT equivalent • Specialized explosive charges (breaching charges, shape charges, ballistic discs, etc.) • Pistols, submachine guns, assault rifles as defensive weapons • Official or corporate uniforms and credentials	• Pedestrian - 30kg/65lbs TNT equivalent • Motorcycle - 45kg/100lbs TNT equivalent • Car - 180kg/400lbs TNT equivalent • Van - 450kg/1000lbs TNT equivalent • Mid-size Truck - 4500kg/10000lbs TNT equivalent • Large Truck - 18000kg/40000lbs TNT equivalent • Official or corporate uniforms and credentials

FIGURE B.13

Port-based scenarios.

Stand-off Attack	Airborne Attack	Labor Unrest	Insider Threat (Attack, Sabotage, Theft of Dangerous Materials, Collusion)
• 1-5 attackers • Knowledge of facility through open source information, observation, and/or insider collusion • Trained in the use of explosives and small arms	• 1-5 attackers • Knowledge of facility through open source information, observation, and/or insider collusion • Trained as a pilot	• Large groups, crowds • Knowledge of facility, operations, layouts, and critical nodes through personal experience	• 1-3 attackers • May coordinate attack with outsiders/terrorists • Knowledge of facility, operations, layouts, and critical nodes through personal experience • May be trained on technical processes and equipment
• Pedestrian • Off-road all-terrain vehicles • Motorcycles • Cars, vans, mid-size trucks • Vehicles may be painted/ marked to appear to be official (e.g. ambulance, police) or company	• Small plane - xx kg, 200km/hr max • Medium plane - 5500kg, 750km/hr max • Large plane - 200,000kg, 850km/hr max • Helicopter - xxkg, 200km/hr max	• Pedestrian • Industry cars, vans, mid-size trucks, large trucks	• Pedestrian • Off-road all-terrain vehicles • Motorcycles • Cars, vans, mid-size trucks, large trucks • May be authorized industry vehicles
• Surface-to-Surface missiles. • Sniper rifles (up to .50 caliber) • Rocket propelled grenades (RPG) • Pistols, submachine guns, assault rifles as defensive weapons	• Small plane* - 180kg/400lbs TNT equivalent • Medium plane* - 450kg/1000lbs TNT equivalent • Large plane* - 4500kg/10000lbs TNT equivalent • Helicopter* - 450kg/1000lbs TNT equivalent	• Access to industrial hand tools but not necessary to cause damage • Official or corporate uniforms and credentials	• Access to sophisticated tools and industrial hand tools but not necessary to cause damage

	Small-Arms Assault	Airborne Attack	Covert Intrusion	Explosive Attack (Boat/Swimmer/Cargo Delivery)
Attackers	• 1-12 attackers • Knowledge of facility through open source information, observation, and/or insider collusion • Trained in the use of explosives and small arms • Willingness to sacrifice part of or entire team	• 1-5 attackers • Knowledge of facility through open source information, observation, and/or insider collusion • Trained in the use of explosives and small arms • Trained as a pilot • Willingness to sacrifice entire team	• 1-5 attackers • Knowledge of facility through open source information, observation, and/or insider collusion • Trained in the use of explosives and small arms • Willingness to sacrifice part of or entire team	• 1-5 attackers • Knowledge of facility through open source information, observation, and/or insider collusion • Trained in the use of explosives and small arms • Willingness to sacrifice part of or entire team
Transport	• Boats • Boats may be painted/ marked to appear to be official (e.g. Coast Guard, police) or company • Could use dhows • Could use more than one vessel	• Small single engine aircraft • Helicopter • Highjacked airliner	• Boats • Swimmers	• Boats • Boats may be painted/ marked to appear to be official (e.g. ambulance, police) or company • Cargo in trucks or maritime conveyance • Dhows which are ubiquitous in the region
Weapons/ Equipment	• Pistols, submachine guns, assault rifles, sniper rifles (up to .50 caliber), light machine guns • Rocket propelled grenades (RPG) • Grenades (high explosive and incendiary) • Bulk explosives for access or attack (up to 4500kg/10000lbs TNT equivalent depending on vehicle) • Specialized explosive charges (breaching charges, shape charges,, etc.)	• Explosives • Knives/Small arms for highjacking • The aircraft itself	• Mechanical breaching tools, quick saws, chainsaws, sledge hammers, hand tools, ladders/ropes/material to cover topguard • Grenades (high explosive and incendiary) • Bulk explosives - 30kg/65lbs TNT equivalent • Specialized explosive charges (breaching charges, shape charges, ballistic discs, etc.) • Pistols, submachine guns, assault rifles as defensive weapons	• Small boat- 1000kg/2250 lbs TNT equivalent • Mid-size boat- 4500kg/10000lbs TNT equivalent • Large Truck/container - 18000kg/40000lbs TNT equivalent • Dhow – 40000kg/88000lbs TNT equivalent

FIGURE B.14

Maritime scenarios.

Stand-Off Attack	Insider Threat (Vessel/Platform Sabotage/Hijacking)	Sea Mines	Vessel Hijacking
• 1-5 attackers • Knowledge of facility through open source information, observation, and/or insider collusion • Trained in the use of explosives and small arms	• 1-3 attackers • May coordinate attack with outsiders/terrorists • Knowledge of facility, operations, layouts, and critical nodes through personal experience • May be trained on technical processes and equipment • Access to sophisticated tools	• Knowledge of key waterways • Likely to be state-sponsored or a hostile state • Likely to be trained in explosives and vessel handling.	• 4-12 attackers • Knowledge of vessel and cargo through open source information, observation, and/or insider collusion • Trained in the use of RPGs and small arms • May be willing to sacrifice themselves depending on motivation.
• Boats	• Already inside as an employee/ crew member	• Medium to large vessels. • May be cargo vessels, offshore supply vessels, or other smaller industrial-type vessels.	• Boats • Boats may be painted/ marked to appear to be official (e.g. Coast Guard, police) or company • Could use more than one vessel
• Surface-to-Surface missiles. • Rocket propelled grenades (RPG) • Large caliber machine guns • Pistols, submachine guns, assault rifles as defensive weapons	• Industrial hand tools • Use highjacked vessel as weapon • Official or corporate uniforms and credentials	• Sea mines manufactured by a nation or improvised sea mines. • May be contact, proximity, or remote detonation.	• Pistols, submachine guns, assault rifles, sniper rifles (up to .50 caliber), light machine guns • Rocket propelled grenades (RPG) • Grenades (high explosive and incendiary) • Official or corporate uniforms and credentials

Tips for Assessing Risk Appetite

INTRODUCTION

Assessing risk appetite or tolerance is a key component of sophisticated risk management primarily because it allows risk managers and those organizations or facilities being assessed to more effectively determine the potential risk treatments that are most appropriate. Despite the importance of determining risk appetite, it is often overlooked by risk managers or those carrying out risk assessments. This omission results in recommendations for treatment that may not meet the needs or desired approaches of clients or organizations. Further, the client or organization being assessed may not understand the concept of risk appetite or may not be able to articulate it, which will require extra effort on the part of the risk-management team.

DEFINING RISK APPETITE

At its most basic, risk appetite is the amount and type of risk an organization is willing to accept. It pervades all areas of risk, whether it involves security, safety, regulatory issues, reputation, finances, or personal considerations. While many executives of organizations or companies are able to make risk decisions based on an informally assessed and intuitive understanding of their own or their organizations' approach to risk, there is rarely a clear definition of risk appetite or a formal process for determining and documenting it.

Essentially, risk tolerance or appetite is the amount of risk that can be accepted by a person or entity without the requirement to treat the risk. An established process to include this in risk management methodologies or approaches will serve to more effectively identify the critical elements of an organization or operation and will likely prevent tendencies to risk aversion.

Risk Appetite and ISO 31000

Most risk-assessment methodologies either do not address or address in a very cursory manner the issue of identifying the risk tolerance of the subject of the risk-management exercise. If it is accepted that risk management is really about managing risk, not just mitigating or reducing risk, the determination of risk tolerance is of vital importance. This process needs to be included in risk assessments so the appropriate and tailored risk-treatment measures can be developed to meet the requirements of the protected entity.

The International Standards Organization Standard 31000 on Risk Management – Principles and Guidelines (ISO 31000) is an internationally accepted approach to risk management. In order to be most effective, however, the standard requires additional focus on the assessment of risk appetite or tolerance, both of which are key elements to the development of a realistic, rigorous, and accurate risk assessment. ISO 31000 recognizes the importance of understanding risk appetite but does not include a description of how the process of determining it should be carried out.

Assessing Risk Appetite

At the outset of a risk-management activity, it is useful to gauge the extent to which decision-makers are prepared to tolerate risk. Understanding their risk appetite facilitates the development of strategies for prioritizing and mitigating risk. Assessing a client's risk tolerance or appetite should be carefully developed and validated throughout the assessment process. Clients who do not have a sophisticated understanding of risk may not be willing or able to articulate their risk tolerance. This creates a challenge for the risk analyst because a higher-end risk assessment and suggested treatments cannot be successfully accomplished without determining risk tolerance.

Helping a Client Determine Risk Appetite

As noted, the client's overall approach to risk may be difficult to ascertain and will depend on his or her level of sophistication regarding risk management. For relatively unsophisticated clients, the initial response is often unclear or may appear to be risk-averse, as the client has not committed to accepting a certain level of risk and is therefore extremely uncomfortable. Further, the client may be influenced by the attitudes of influential stakeholders, who may either be inexperienced in determining risk appetite or unwilling to support any acceptance of risk. For these reasons, this is a sensitive yet vital issue that needs to be introduced carefully to clients and stakeholders who are unfamiliar with the concept or process.

The most common methods to ascertain risk tolerance can include workshops, questionnaires, and stakeholder interviews. However, in cases where the client is reluctant to articulate risk tolerance, it is incumbent on the consultant to develop a potential risk-tolerance model based on stakeholder and client engagement during the course of the assessment. This can be done by performing or collecting information from the following:

- Conducting a risk-appetite presentation to key client and stakeholder representatives
- Carrying out workshops and interviews with key stakeholders
- Asking stakeholders to answer a tailored questionnaire on risk appetite
- Obtaining and analyzing existing security assessments for indications of what constitutes a major incident as well as those functions that have been identified as critical
- Obtaining and analyzing crisis and emergency plans for response triggers and measures that may indicate the level of importance given to various events and potential risks
- Interviewing and reviewing documents and assessments from the enterprise risk- management team, if extant

- Reviewing business-continuity plans, including impact analysis reports to identify organizational criticalities and recovery-time objectives (RTOs)

When the data collection is completed, it should be analyzed and a report generated with several options regarding risk appetite. The data will provide a more focused understanding of the organization's critical functions and hopefully a basic understanding of the level of the client's and stakeholders' risk appetite or level of risk aversion. Categories analyzed should at a minimum include potential human losses, monetary losses, reputational effects, and the losses of critical functions at varying levels.

After preparing an initial report with optional levels of risk appetite identified, it may be useful to refine these findings and gain client and stakeholder validation of the assessment by engaging in a "pairwise" exercise with clients.

Pairwise Exercise

Pairwise comparisons are based on the idea that two similar options or "things" are presented to an audience, and the audience is asked to state which "thing" is preferred. This is particularly useful when the audience is initially unsure of which option or "thing" is preferred or if choices are so varied that there needs to be a process for narrowing them down. A common example of a pairwise exercise is found in eye examinations. The optometrist or technician will make a general assessment of the basic prescription that is likely to be most accurate for the patient and will then use a machine to show the patient pictures using different lenses. The patient will be asked whether option 1 or option 2 is clearer. This process refines the prescription by allowing the patient to compare fairly similar lenses for comfort and clarity.

By using a pairwise approach, the overwhelming and uncomfortable nature of assessing risk appetite can be reduced or eliminated by allowing stakeholders and clients to compare specific criteria against clearly defined critical functions or key areas of importance. This process can lead to an accurate assessment of risk appetite. It is important to note that the risk analyst needs to be careful not to steer the pairwise exercise to a desired outcome and to use options derived from a rigorous assessment based on the sources noted previously.

Risk Appetite and Risk Treatment

Upon completion and validation of the risk-appetite assessment, which is carried out concurrently as part of the risk assessment, the findings should be factored into the risk register and its relative rankings.

As a result, the relative ranking of risks will enable decision-makers to decide on the proper risk treatment for the risks identified and ranked as most important. This involves selecting one or more options for addressing those risks in accordance with the agreed-upon risk-tolerance analysis that informs the risk register.

An effective way to include the risk-appetite findings in a risk register is to include the results of the risk-appetite analysis in the consequence ratings and ensure that they have been validated by the appropriate stakeholders and the client.

The risk appetite should also be an essential part of the consideration of risk treatments. In treating risks, decision-makers can consider a number of options, either in combination or independently:

- Accepting the risk by not implementing any countermeasures
- Avoiding the risk by discontinuing the activity that presents a risk or instituting measures that mitigate threat, vulnerability, or consequence
- Reducing risk by putting in place risk-management measures; this is the most common approach when a fully developed risk-management program does not exist
- Transferring risk to another entity such as an insurance company—this involves the recognition that the identified risk is too significant to be avoided or accepted but it cannot be mitigated

Ultimately, it is incumbent on the client to evaluate the respective costs and benefits of each risk-mitigation investment in order to determine the most effective for their particular jurisdiction and to make final decisions, based on the best advice provided by the risk analyst. However, the assessment of risk appetite is an essential component of this process, and the development of risk-treatment options is not easily defensible if a formal process to determine risk appetite or tolerance is lacking. The ability to perform a risk-appetite assessment coincident with the risk assessment and incorporate it into the risk-treatment strategy is a critical and generally overlooked component of a comprehensive and sophisticated approach to risk management, especially in a complex operating environment such as the maritime domain and international shipping.

Survey on Risk Appetite
Date:
Representative Name and Title:
Contact Information **(Phone/E-mail):**

Which of the following statements best describe your experience with risk management?

☐ No experience
☐ Limited experience
☐ Experience with financial risk management
☐ Experience with corporate or business risk management
☐ Experience with physical security risk management

Which of the following statements is important to you when considering risk management?

	Not important				Very important
1. To avoid risk of any sort	A	B	C	D	E
2. To seek options to transfer risk to others	A	B	C	D	E
3. To offset potential impact of risk	A	B	C	D	E
4. To prepare a comprehensive strategy	A	B	C	D	E
5. To address all foreseeable risk	A	B	C	D	E

Which of the following risk criteria is important to you?

	Not important				Very important
1. Organizational output (time, cost, quality)	A	B	C	D	E
2. Resources	A	B	C	D	E
3. Reputation	A	B	C	D	E
4. Business continuity	A	B	C	D	E
5. Clients/stakeholders	A	B	C	D	E
6. Compliance with government strategy/policy	A	B	C	D	E

In financial terms, what do you consider to be a "moderate" financial loss?

☐ 1,000 (USD)
☐ 10,000
☐ 100,000
☐ 1,000,000
☐ 10,000,000

As a manager within your organization, when would you require a briefing from your staff in relation to a security incident?

☐ After any incident, regardless of how insignificant
☐ After any minor incident
☐ Only if the nature of the incident has at least a moderate impact
☐ Only if the nature of the incident has a major impact on your business
☐ Only if the nature of the incident has the potential for a catastrophic impact on your business

FIGURE C.1   *(Figure C.1 continues on next page)*

Generic model of risk appetite.

Do you agree with the following statements:

Severe risk must be avoided under all circumstances	Y/N
High risk must be mitigated and constantly monitored	Y/N
Moderate risk should be managed and reduction strategies implemented	Y/N
Low risk may be acceptable after a review	Y/N
Very low risk would normally not be treated but monitored	Y/N

Any other comments regarding levels of acceptable risk for your organization or operations?

When considering the likelihood of an undesirable event occurring, what timeframe are you most concerned with?

1. Monthly
2. Quarterly (3 months)
3. Yearly
4. 2–3 years
5. 3–5 years

What are your organization's critical functions?

What functions are not critical?

What critical external dependencies that are needed to continue operations have you identified?

Do you agree that the following is a primary source of threat/hazard against your organization or operations?

1. Criminal
2. Terrorism
3. State entity
4. Industry competition
5. Staff or former staff
6. Acts of nature
7. Accidents
8. Lack of training/oversight
9. Other (explain)

FIGURE C.1 *(continued)*

Which of the following is important to you when considering influencing factors that contribute to risk?					
	Not important			Very important	
Culture	A	B	C	D	E
Internal stakeholders	A	B	C	D	E
External stakeholders	A	B	C	D	E
Organizational structure	A	B	C	D	E
Business type	A	B	C	D	E
Which of the following is important to you when considering the impact of risk upon goals and objectives?					
	Not important			Very important	
1. Culture	A	B	C	D	E
2. Internal stakeholders	A	B	C	D	E
3. External stakeholders	A	B	C	D	E
4. Organizational structure	A	B	C	D	E
Which of the following is important to you when considering the potential implications of program failure?					
	Not important			Very important	
1. Culture	A	B	C	D	E
2. Internal stakeholders	A	B	C	D	E
3. External stakeholders	A	B	C	D	E
4. Organizational structure	A	B	C	D	E

FIGURE C.1 *(continued)*

Index

Printed in the United States
By Bookmasters